T0174215

"Current discussions about transitions are often dominated by techno-economic perspectives that obsess over elements like carbon and cost. This book takes a refreshing departure from that paradigm, and it explores the cognitive, psychological and behavioural elements underlying those very same transitions. It underscores that if we truly want a more sustainable future, we need to change not only the technologies we build and the market mechanisms we design, but how we *think* about ourselves, society, and the sociotechnical systems embedded between them all."

Benjamin K. Sovacool, *Professor of Energy Policy at the Science Policy Research Unit (SPRU) at the School of Business, Management, and Economics, University of Sussex, UK*

"Transition studies and (environmental and social) psychology have strongly contributed to our understanding of sustainability transitions. However, and surprisingly, the streams of literature have so far developed in relatively unconnected ways. I thus consider this book a valuable addition to the literature."

Elisabeth Dütschke, *Senior Scientist at Fraunhofer Institute for Systems and Innovation Research (ISI), Germany*

"This timely book is a must-read for all transition scholars interested in studying agency in sustainability transitions. It is also a welcome invitation for a greater engagement of social psychologists in studying the role of individuals within larger socio-technical transition processes towards sustainability. May it inspire new interdisciplinary research leading to an enhanced understanding of how to accelerate sustainability transitions in sectors such as energy, mobility or agro-food."

Karoline Rogge, *Senior Lecturer at the Science Policy Research Unit (SPRU) and Co-Director of the Sussex Energy Group, University of Sussex, UK, and Senior Researcher, Fraunhofer ISI, Germany*

"This is an important book for social scientists and others concerned with sociotechnical transitions. The authors have identified and explored a hugely significant aspect of energy transitions and energy technology diffusion, acceptance and use: the connection between individual-level psychology and sociotechnical processes. The message is very relevant: social psychology is not only important to sociotechnical transitions in the context of energy supply and use, but that the differences in analytic levels are bridgeable."

Christian Oltra, *PhD, Senior Researcher at the Centro de Investigación Socio-Técnica, CIEMAT Barcelona*

Energy Transitions and Social Psychology

This book explains how social psychological concepts can be closely integrated with sociotechnical perspectives of energy transitions. It shows the value of actor-centred analysis that acknowledges the role of individual-level processes within their wider contexts of energy supply and use. In this way, the book connects social psychological and sociological frames of analysis, preserving the value of both, to provide multi-level, analytically extended accounts of energy transitions processes.

Sociotechnical thinking is about the interactions of people and technology, including the rules, regulations and institutions involved. Such perspectives help to identify the many forms of path dependency that can make change difficult. Human behaviour plays a strong role in maintaining these path dependencies, but it can also introduce change. This book advocates a deliberately interdisciplinary research agenda that recognises the value of social psychological perspectives when seeking to create new pathways for energy supply and use. At the same time, it also demonstrates the value of sociotechnical perspectives for energy-related social psychology.

Energy Transitions and Social Psychology will be of great interest to students and scholars of energy transitions, environmental and energy psychology, sustainable development and innovation studies, as well as students and scholars of environment and energy more generally.

Paul Upham is Chair of Human Behaviour and Sustainable Development at Leuphana University, Germany, and Visiting Professor at the Copernicus Institute of Sustainable Development at Utrecht University, the Netherlands.

Paula Bögel is a post-doctoral researcher in the social psychology of energy transitions at the Chair for Human Behaviour and Sustainable Development at Leuphana University, Germany and of the research group Urban Analytics and Transitions at the Royal Institute of Technology (KTH), Stockholm, Sweden.

Katinka Johansen is a post-doctoral researcher in energy social science at the Chair for Human Behaviour and Sustainable Development at Leuphana University, and has recently completed a PhD in social psychology at the Technical University of Denmark.

Routledge Studies in Energy Transitions
Series Editor: Dr. Kathleen Araújo
Boise State University and Energy Policy Institute,
Center for Advanced Energy Studies (US)

Considerable interest exists today in energy transitions. Whether one looks at diverse efforts to decarbonize, or strategies to improve the access levels, security and innovation in energy systems, one finds that change in energy systems is a prime priority.

Routledge Studies in Energy Transitions aims to advance the thinking which underlies these efforts. The series connects distinct lines of inquiry from planning and policy, engineering and the natural sciences, history of technology, STS, and management. In doing so, it provides primary references that function like a set of international, technical meetings. Single and co-authored monographs are welcome, as well as edited volumes relating to themes, like resilience and system risk.

Series Advisory Board
Morgan Bazilian, Colorado School of Mines (US)
Thomas Birkland, North Carolina State University (US)
Aleh Cherp, Central European University (CEU, Budapest) and Lund University (Sweden)
Mohamed El-Ashry, UN Foundation
Jose Goldemberg, Universidade de Sao Paolo (Brasil) and UN Development Program, World Energy Assessment
Michael Howlett, Simon Fraser University (Canada)
Jon Ingimarsson, Landsvirkjun, National Power Company (Iceland)
Michael Jefferson, ESCP Europe Business School
Jessica Jewell, IIASA (Austria)
Florian Kern, Institut für Ökologische Wirtschaftsforschung (Germany)
Derk Loorbach, DRIFT (Netherlands)
Jochen Markard, ETH (Switzerland)
Nabojsa Nakicenovic, IIASA (Austria)
Martin Pasqualetti, Arizona State University, School of Geographical Sciences and Urban Planning (US)
Mark Radka, UN Environment Programme, Energy, Climate, and Technology
Rob Raven, Utrecht University (Netherlands)
Roberto Schaeffer, Universidade Federal do Rio de Janeiro, Energy Planning Program, COPPE (Brasil)
Miranda Schreurs, Technische Universität München, Bavarian School of Public Policy (Germany)
Vaclav Smil, University of Manitoba and Royal Society of Canada (Canada)
Benjamin Sovacool, Science Policy Research Unit (SPRU), University of Sussex (UK)

Titles in this series include:

Visions of Energy Futures
Imagining and Innovating Low-Carbon Transitions
Benjamin K. Sovacool

Energy Transitions and Social Psychology
A Sociotechnical Perspective
Paul Upham, Paula Bögel and Katinka Johansen

Narratives of Low-Carbon Transitions
Understanding Risks and Uncertainties
Edited by Susanne Hanger-Kopp, Jenny Lieu and Alexandros Nikas

For more information about this series, please visit: www.routledge.com/Routledge-Studies-in-Energy-Transitions/book-series/RSENT

Energy Transitions and Social Psychology

A Sociotechnical Perspective

**Paul Upham, Paula Bögel
and Katinka Johansen**

Routledge
Taylor & Francis Group

LONDON AND NEW YORK

from Routledge

First published 2019
by Routledge
2 Park Square, Milton Park, Abingdon, Oxon OX14 4RN

and by Routledge
52 Vanderbilt Avenue, New York, NY 10017

First issued in paperback 2020

Routledge is an imprint of the Taylor & Francis Group, an informa business

© 2019 Paul Upham, Paula Bögel and Katinka Johansen

The right of Paul Upham, Paula Bögel and Katinka Johansen to be identified as authors of this work has been asserted by them in accordance with sections 77 and 78 of the Copyright, Designs and Patents Act 1988.

All rights reserved. No part of this book may be reprinted or reproduced or utilised in any form or by any electronic, mechanical, or other means, now known or hereafter invented, including photocopying and recording, or in any information storage or retrieval system, without permission in writing from the publishers.

Trademark notice: Product or corporate names may be trademarks or registered trademarks, and are used only for identification and explanation without intent to infringe.

British Library Cataloguing-in-Publication Data
A catalogue record for this book is available from the British Library

Library of Congress Cataloging-in-Publication Data
A catalog record has been requested for this book

ISBN 13: 978-0-367-66146-5 (pbk)
ISBN 13: 978-1-138-31175-6 (hbk)

Typeset in Sabon
by Wearset Ltd, Boldon, Tyne and Wear

Contents

Figures

Tables

Boxes

Author biographies

Paula Bögel is a post-doctoral researcher in the social psychology of energy transitions at the Chair for Human Behaviour and Sustainable Development at Leuphana University, Lüneburg, Germany and of the research group 'Urban Analytics and Transitions' at the Royal Institute of Technology (KTH), Stockholm. With a background in business studies, psychology and sustainability sciences, Paula studies the dynamics of sociotechnical sustainability transitions. She aims to contribute to transition studies research by improving our knowledge about agents' psychology in transitions. Her studies on human behaviour in sociotechnical transitions – be these in relation to consumers, employees, experts, politicians or the general public – aim to help provide an understanding of why agents react in the way they do and how these behaviours can be transformed to improve social acceptance of new, more sustainable technologies and sustainable development.

Katinka Johansen holds a PhD in social psychology and an MSc in social anthropology, with research interests across energy transitions, environmental governance and research design. Her PhD focuses on public perceptions of renewable energy technologies, specifically offshore wind energy technologies, and it touches upon public, political and environmental perspectives in that regard. Her professional toolkit comprises both qualitative, quantitative and mixed methods research approaches, and she has conducted fieldwork in Bolivia, Northern Uganda and Denmark.

Paul Upham PhD is Chair of Human Behaviour and Sustainable Development at Leuphana University, Lüneburg, Germany and Visiting Professor at the Copernicus Institute of Sustainable Development at Utrecht University, the Netherlands. He was previously a Senior Research Fellow at the Sustainability Research Institute, University of Leeds, UK; Senior Research Fellow at Manchester Institute of Innovation Research and Tyndall Centre for Climate Change Research at the University of Manchester; and Visiting Professor in Governance of Energy Systems and Climate Change at the Finnish Environment Institute (SYKE), Helsinki. With a psychology background, Paul works on public and stakeholder opinion and engagement in energy technology governance and sociotechnical transitions.

Preface

A book about the psychology of energy transitions, particularly sociotechnical sustainability transitions, in the mode that has coalesced around Frank Geels' (2002) multi-level perspective (MLP), needs some explanation. This is particularly because this book is not – as we reiterate in the introduction – about simply showing the *relevance* of psychology for energy transitions. Rather, the book is also about how human behaviour- and agency-related social science perspectives with different underlying ontologies may be *brought together more closely* at a theoretical level, and it is about the value of doing so.

Outside of academia, an emphasis on theoretical integration may seem abstruse and somewhat irrelevant. However, while the sociotechnical approach to sustainability transitions thinking has grown in extent and influence, disciplinary specialisation has (we think) left the psychology of the actors involved under-examined. This book is intended in part as a response to this, specifically in the context of energy transitions, although with wider relevance. The book is equally a response to our frustration with social psychological analysis of energy-related psychology that gives too little attention to sociotechnical context. Workable solutions to increasingly pressing sustainability and, specifically, energy problems need to address interrelated phenomena in coherent, reinforcing ways. This requires different types of knowledge and understanding and as such is starkly at odds with academic specialisation, despite the latter having its own value.

The book coheres ideas that we have been developing, with colleagues, for several years. It is intended to both set energy-related social psychology in the wider context that sociotechnical sustainability transitions offer, and to draw attention to the value of social psychology for the sustainability transitions literature. We hope to stimulate some thought on these matters – whether or not you agree with what we are doing, or how we are doing it. We are well aware that the book only begins to address the contribution that psychology can make to understanding individual-level processes in energy transitions, and that there are many more psychological perspectives that can be drawn upon that we have not had the time to use here. Hopefully, there will be a lot more such work in the years to come.

Paul Upham, Paula Bögel and Katinka Johansen

Reference

Geels, F. W. 2002. "Technological transitions as evolutionary reconfiguration processes: A multi-level perspective and a case-study". *Research Policy* 31:1257–74.

Acknowledgements

This book draws on a programme of energy social science undertaken over several years, which we reference throughout. For further detail on methods, for example, please see the publicly available pre-print versions of our co-authored papers.[1] Many colleagues were involved to varying extents in data collection, analysis, discussion and writing of the underlying, contributory work. These include the following people (in alphabetical order): Dr Sophia Becker (Institute for Advanced Sustainability Studies, Potsdam); Dr Elizabeth Dütschke (Fraunhofer Institute for Systems and Innovation Research); Professor Mikael Hildén (SYKE, the Finnish Environment Institute, Helsinki); Dr Rita Klapper (Copernicus Institute of Sustainable Development, Utrecht University); Dr Paula Kivimaa (SYKE and SPRU, the Science Policy Research Unit, University of Sussex); Dr Les Levidow (Open University, UK); Monica Lores (CIEMAT, the Centre for Energy, Environment and Technology, Barcelona); Dr Chris Martin (University of Leeds); Christian Oltra (CIEMAT); Uta Schneider (Fraunhofer Institute for Systems and Innovation Research); Dr Rosa Sala (CIEMAT); Venla Virkamäki (ex-SYKE); Dr Zia Wadud (Institute for Transport Studies and Centre for Integrated Energy Research, University of Leeds). Of course, responsibility for the content of the book lies with the authors.

Note

1 www.researchgate.net/profile/Paul_Upham.

Acknowledgements

Part I

Rationale

1 Introduction

Scope and purpose of the book

Decarbonising energy systems, while at the same time ensuring stable, sustainable and affordable energy supplies, is likely to have major ramifications for the publics[1] who are asked to accept these new energy infrastructures and technologies. In a future with more energy storage and intermittent renewable supply, publics are also likely to be nudged towards accepting some degree of change in their energy consumption patterns (Jacobsson and Johnson 2000). Hence public opinion and practices have become matters of importance for governments, the energy industry and academics alike (Devine-Wright 2009; van der Horst and Toke 2010). In particular, the way in which some renewable (and non-renewable) energy infrastructure projects have faced opposition from the local communities where they are constructed, while others have not (Toke 2005), has contributed to the growing interest in understanding the factors that drive public reactions to energy technology projects. Indeed, it would be fair to say that "social acceptance" of energy technologies has become one among many policy relevant social science concepts in the field of energy transition studies (Sovacool 2014).

The main aim of this short book is to show how concepts that help to explain individual-level psychology can be integrated with sociotechnical transitions theory, with a particular focus on *energy*. The focus here on integration is a more specific and challenging aim than simply showing the relevance of psychology in this context. Sociotechnical transitions theory – which we explain in more depth below – is most often framed in terms of different collectives of system actors, rather than in terms of individual people. Hence, the literature deals with the roles of firms, agencies, organisations as intermediaries, institutions and, less commonly, with entrepreneurs (policy, for profit and non-profit). Moreover, the sociotechnical transitions literature also seeks to generalise from individual cases, to suggest system-level patterns based on observation of the past, which can inform measures to accelerate energy and other transitions in the future.

All of this mitigates against – and is distinct from – analysis of intra-individual and social psychological processes (see Box 1.1 for an introduction to social psychology). Nonetheless, here we want to show why (social) psychological processes are important for transitions and how these individual-level processes can be analysed in relation to sociotechnical frameworks. While it is arguably straightforward to show the relevance of psychology to sociotechnical change in energy systems via the logical connection of ideas – for example, to show that consumer or citizen psychology has implications for technology use or acceptance and hence for system change – it requires more consideration to find psychological concepts that span individual and social levels, and to find theoretical approaches and methods that support such integration. In this book we use the terms 'social psychology' and 'psychology' interchangeably for brevity.

Box 1.1 Social psychology

In 1954, Gordon Allport defined social psychology as "the scientific attempt to explain how the thoughts, feelings and behaviors of individuals are influenced by the actual, imagined, or implied presence of other human beings" (Allport 1954, p. 5). Social psychology is thus a broad research domain covering multiple research topics. Overall, social psychologists are concerned with individual behaviours. Although they are specifically concerned with the behaviour of individuals in a broader social context, that context is often narrowed down to a number of specific terms (indeed, we might sometimes say overly-specific).

Here, *thoughts* refer to the beliefs or expectations that people/actors may have vis-à-vis particular phenomena in the world; *feelings* refer to the broad range of emotional reactions, moods and motivations that people experience; and *behaviours* refer to observable actions, i.e. the 'things' that we see people doing. Overall, social psychologists share an interest in groups and societies, but their focus is on the individuals that are a part of those groups or societies. Social psychologists seek an understanding of individual persons; uncovering what they care about, how they operate, what their motivations are, and social psychologists do so with the aim of understanding how these individuals navigate their social environments. Of key importance is how people and groups both influence and are influenced by other human beings.

Social psychologists investigate a range of human experiences, for example: intrapersonal phenomena (occurring within the self or in the mind), such as cognition, perception, self-concepts and the behaviours we enact, and interpersonal phenomena (occurring between people), such as social influence and group dynamics. For example, the way we feel about ourselves, e.g. our self-esteem, may mirror how we fit into particular social groups in our social environment. And the way we see and perceive the world, e.g. the way we see other people or peoples, is greatly shaped by influence and persuasion from our peers and the multiple other sources of information we are subjected to in our lives. Thus, our being in the world is highly social, and it is precisely this social psychological material that is the core interest of social psychologists.

Historically, sociology and psychology have been closely related, with social psychology to some extent linking the two disciplines. While sociology and psychology have become progressively more specialised as disciplines, arguably social psychology in general still does cover some common ground between the two. Reflecting the broad nature of social psychology, multiple types of research methods, addressing the very different types of research inquiry within the field, have been used. Social psychology is an empirical science in that it investigates and reasons from observable phenomena to find answers regarding human behaviours. Diverse methods and research designs are used, from strictly controlled psychological experimental studies to correlational studies, where statistical methods are used to identify contribution to variance, to more qualitative and observational methods. While quantitative methods have perhaps dominated in sustainability-related psychological research, observational and case-based qualitative methods also provide data that can be readily integrated into the types of case-based research design historically favoured in sociotechnical transitions studies.

In summary: for its vast and very diverse array of past and potential future research topics, social psychology may be commonly described as attentive towards the social components of individual behaviours. Social psychology is an empirical science; it is about individual people; it deals with our thoughts, feelings and behaviours; it uses and infers from many types of data; and it does so mindful of how all of these phenomena are shaped by the social world around us (Ross *et al.* 2010; Stangor 2014).

For these reasons, while we have tried to write this book in an accessible style, it is somewhat theoretical in its leaning, partly because it addresses some of the issues that underlie the relative absence of psychology in the energy-oriented sociotechnical transitions literature. That said, we hope the book is still comprehensible by students at different levels, even if some of the issues discussed in Part I may be beyond what they need to deal with.

The book is divided into two parts. Part I deals with the somewhat abstract issues of bridging ontological and disciplinary differences in studies of human behaviour in relation to energy transitions. Here, we set out an integrative approach to connecting the psychological with the sociotechnical. We also define various terms and perspectives that may be unfamiliar to some readers. Part II illustrates how close connections between psychological and sociotechnical concepts can be made in the context of energy transitions, despite differing underlying ontologies. As we expect that there will also be interest in the more general relevance of psychology to energy transitions, in Chapter 9 we set out a number of possibilities and research directions, while leaving more in-depth investigation of connections to systems-level transitions processes for future work.

Sociotechnical transitions perspectives on unsustainability

In this book we emphasise sociotechnical perspectives on energy trans-itions. These perspectives seek to integrate a wide range of relevant factors with broad, system-level descriptions. What they lose in terms of detail, they gain in breadth of application. Moreover, these perspectives arguably capture much of our current, problematic situation. Despite progress in raising standards of living and quality of life for many, at a global level we are continuing on unsustainable pathways and are far from achieving inter- and intra-generationally equitable development (Rockstrom 2009; Figueres *et al.* 2017). Globally, greenhouse gas emissions continue to increase; bio-diversity loss accelerates; global poverty reduction fails to meet United Nations goals; social inequality is intensifying and economic instability threatens societal cohesion and political stability (Heinrichs *et al.* 2015). While sustainability awareness grows internationally, as reflected in global policy initiatives such as the 2030 Agenda (United Nations General Assembly 2015) and the Paris Agreement (UNFCCC 2015), the combined actions of states, companies and civil society around the world that are intended to be mitigative have not – as yet – reversed the unsustainable dynamics of contemporary systems of global production and consumption of all commodities *at a global net level*.

Why is this? From a sociotechnical transitions perspective, technological development, science, industry, markets, policy and culture are inter-twined, co-evolving in a complex system of mutual and often self-reinforcing and supporting interactions (Kemp *et al.* 1998; Geels and Schot 2007). Successful intervention requires an understanding and anticipation of the connections and, hence, possible consequences. Moreover, successful intervention is inherently political (Smith *et al.* 2005): what constitutes success for one set of interests often implies loss for another. Indeed, despite the systems and managerial discourse of 'transition management' – i.e. the attempt to deliberately intervene and to steer transition processes – such practice is inherently political and seeks to change various forms of power (Avelino and Rotmans 2009).

Most sociotechnical transitions theorists study historical and con-temporary cases of co-evolving social and technological change and stasis in the search for ways of accelerating and managing sociotechnical traject-ories that are more sustainable, with particular attention given to describ-ing and explaining change and resistance to change (Ulli-beer 2013). One of the most commonly-used conceptual models of sociotechnical systems is Geels' (2002) multi-level perspective (MLP). This model proposes that sociotechnical systems can be usefully analysed according to three different levels of practice: the niche, regime and landscape (Geels 2002), as explained in Box 1.2. Understanding regime-level dynamics has proved to be particularly important when it comes to understanding the lack of change and resistance to change in established sociotechnical systems

(Geels 2014). We will refer to the MLP heuristic or framework frequently during the course of the book.

Box 1.2 The multi-level perspective

Energy-focused, sociotechnical sustainability transition studies have focused on pathways and processes that help to explain interconnected social and energy technology change processes, from the development to the deployment stage. For this purpose, the multi-level perspective (MLP) has proved to be a popular framework (Geels 2002, 2011; Schot and Geels 2008).

As stated above, the MLP assumes three structural levels of analysis: the landscape, the regime and the niche (Geels 2002; Geels and Schot 2007). *Niche* innovations are bottom-up phenomena; niches are protected spaces allowing for the experimentation with emergent technologies, user practices and regulatory structures (Kemp *et al.* 1998; Schot and Geels 2008). The *regime* is the dominant set of ideas and ways of doing things, which are institutionalised in all senses. At the regime level, strong cultural influences and the effects of macro-political dynamics and economics lead to and reflect relative stability and often resistance to change that threatens established interests. Here, economic actors respond to the market signals operating within the technological paradigms or regimes (Van den Belt and Rip 1987; Van Den Ende and Kemp 1999). The regime includes meso-level dominant technologies, institutions, practices and rules. Regime level dynamics have proved to be important for understanding the lack of change and/or resistance to change in established sociotechnical systems.

The *landscape* is theoretically located at the macro-level of this three-level view of the world. It responds to and consists of multiple processes, including political, economic, social demographic and cultural, but it is above all a 'place' of slow change and, to some extent, 'taken-for-grantedness'. Thus, cultural values and slow-changing technological, structural and social change processes are conceptually located in the landscape (Geels 2002, 2014). It might be said that the landscape is the least theorised and analysed of the levels in the sociotechnical sustainability transitions literature.

Overall, the MLP holds that transitions processes are nonlinear, and that they are the result of societal and sociotechnical systems interactions at multiple levels. The regime is under pressure from changes at the level of the niche and the landscape, and sometimes this leads to regime destabilisation. Ultimately, when niches mature, regimes destabilise and landscapes change, a shift in sociotechnical regimes may be the consequence (Geels 2002; Geels and Schot 2007).

Why does sociotechnical sustainability transitions theory and practice need social psychology?

Some psychological ideas are already implicitly present in sociotechnical sustainability transitions theory. In the innovation studies literature from which sociotechnical transitions perspectives have developed, psychological processes are acknowledged as important in 'selection environments', which are the contexts in which firms (or other actors) make investment decisions. From an evolutionary economic perspective (Nelson and Winter 1982), Dosi (1982) argued that technological paradigms strongly condition selection environments and hence the direction of technological change. Technological paradigms – like the idea of scientific paradigms, from which the parallel was drawn – are a set of consistent and mutually supportive ideas about the nature of a technological problem and how the problem can be solved (Dosi 1982). This is partly why most bicycles have two wheels and look similar, as do most smartphones, or laptops or knives or whatever. Implicit within the idea of a paradigm is the idea of paradigm change – the recumbent bicycle, for example, or the smart watch that performs the functions of a smartphone. The idea of the *technological regime* is close to that of the technological paradigm, and this is where psychology plays a role: cognitive routines are viewed as the key mechanisms by which such paradigms or regimes are maintained. In the sociotechnical world, ideas of how things are – and should be – shape *what is*. Clearly, some ideas have more powerful interests behind them than others, and some ideas have more legitimacy with particular firms, citizen groups, consumers or governments: not all ideas are equal in their capacity to influence. We discuss this later in relation to expectations of energy technologies (see Chapter 5).

Nonetheless, while the routines and heuristics that underlie the original idea of the regime are still a part of sociotechnical transitions accounts, and while they are acknowledged as important, psychological processes in general are given little attention relative to broader, collective processes (Bögel and Upham 2018). Indeed, with its emphasis on systems-level processes, sociotechnical transitions thinking has been critiqued for generally paying less attention to individual agents, despite individual agency having long been recognised as important for transitions processes (Smith *et al.* 2005; Geels 2011).

Transition researchers do implicitly highlight the role of subjective human experience when referring, for example, to the roles of meanings, interpretation, discourses and symbols (Stedman 2016). Our premise, though, is that psychological explanations of various aspects of individual agency have their own intrinsic value; that while there may be sociological analogues or equivalents for understanding particular processes "ignoring insights from psychological research can handicap progress towards a low-carbon, sustainable future" (Clayton *et al.* 2015). We suggest that it is the emphasis on various forms of collective agency that has led to the sociotechnical transitions literature giving less attention to psycho-social processes (see Smith *et al.* 2005; Hynes 2016). Collective agency expressed through institutions and organisations lends itself to explanatory accounts that involve shared, social processes. Yet accounts of individual-level processes also have much to offer in terms of understanding the behaviour of individuals. In Chapter 3 we show that while there *are* a small number of studies that refer to psychological perspectives in the sociotechnical transitions literature, these use a rather limited number of psychological theories, and mostly from a functional perspective (i.e. in terms of outcomes), without examining psychological processes themselves in depth (Nye *et al.* 2009; Whitmarsh 2012; Gazheli *et al.* 2015; Stephenson *et al.* 2015).

While sociological or cultural accounts of subjectively experienced phenomena place their focus external to the individual in terms of processes and emphases, psychology emphasises the characteristics and processes of individuals (micro-level) or groups of individuals *at the level of the individual*. Social psychological accounts do not deny that social processes exist, but their focus is primarily on the processes as they are experienced by the individual. Similarly, sociological accounts generally do not deny individuality, but their primary object is the relationship between the individual and their social context. These distinctions are, of course, only tendencies.

We still haven't quite answered our own question: Why does the sociotechnical sustainability transitions literature, when applied to energy and to other sectors, need psychology? In response, we might equally ask: How could it not? A case against the relevance of psychology would require the argument either that intra-individual processes are irrelevant to energy and other transitions processes, or that sociotechnical transitions processes are conceptualised at such an aggregated scale, and over such extended periods of time, that again individual-level processes have little relevance. Yet the widespread use of detailed case studies and the general attendance to micro-level phenomena and activities make it clear that the sociotechnical, sustainability and energy transitions literature *is* concerned with multi-scalar activity, and that it is also very much concerned with the connections between scales – particularly niche–regime connections, but also regime–regime interactions and incumbent-level (i.e. intra-regime) processes.[2] The purpose of this book is to show why and how individual-level processes merit closer attention as part of these analyses, and not only in consumption and technology acceptance contexts.

Ontological thoughts

When we make assumptions about what people are, or what social structure is and how it comes about, we are making assumptions or assertions about their nature or ontology. Ontological assumptions may also be made for analytic purposes – for example, the MLP divides reality into the three levels of niche, regime and landscape in order to provide a simplified way of thinking about what we know to be complex systems. Nonetheless, despite knowing this, such ways of thinking shape the way in which we see the world or particular aspects of it, functioning as a cognitive heuristic that draws attention to and highlights particular aspects of reality, while downplaying other aspects.

While ontology is perhaps a little abstruse for this early in the book, it is difficult to avoid when seeking to bring together perspectives that are radically different. As mentioned earlier, much sociotechnical thinking makes appeals either to the MLP specifically, or uses the same component concepts. The MLP takes a particular approach to social structure and this has implications for how psychological perspectives might be connected to the MLP, or to thinking that implicitly makes similar assumptions.

For those unfamiliar with any of this, the above probably still sounds cryptic. By way of further explanation, Giddens' (1984) theory of *agency-structure as practice* constituted a seminal insight into the relationship between agency and structure via three main elements or concepts: signification, relating to discourse; legitimation, relating to values, norms and standards; and domination, relating to control over resources. From Giddens' *structuration* perspective, social structures exist and function through agents' actions and agents assign specific roles and meanings to those structures (Grin *et al.* 2010, p. 233). Geels' MLP (2002) uses this idea, assigning different levels of structuration to its three levels. Hence, the niche is least structurated (one might say socially embedded or institutionalised); the material aspects of regime are the outcomes of the structurated social rules that are widely prevalent; and the landscape is at a taken-for-granted level of structuration.

Yet structuration theory has been critiqued for its relatively abstract nature, in particular for neglecting the situated detail of actors' situations, their psychology (such as motivations, attitudes, beliefs and knowledge) and their social, cultural and political contexts (including organisational positions and roles) (Stones 2005). Stones (2005) proposes a solution to this that retains the ideas of structuration and that we expand on and apply. However, another option (which has some similarities in its practical implications for how one might study multi-level phenomena in their own terms), is proposed by Svensson and Nikoleris (2018) and by Sorrel (2018), who both set out similar (and apparently simultaneous) critiques of the ontology of the MLP, advocating a critical realist approach (Bhaskar 1975/1997; Sayer 1992) in favour of structuration.

As Sorrell (2018) observes, the MLP has come to epitomise – or to be taken as epitomising – some of the core concepts of sociotechnical transitions thinking for sustainability. As such, an increasing number of authors are bringing their own disciplinary contributions to the MLP party (so to speak) and this raises issues regarding conceptual and theoretical compatibility and ontological limitations (Sorrell 2018; Svensson and Nikoleris 2018), as well as the explanatory utility of what might be viewed as an ever-expanding theory (Sorrell 2018). These authors advocate critical realism as an (arguably) open-minded perspective that invites different types of explanation, allowing for multiple layers of structure and process that can be analysed in different ways and from the perspective of different disciplines, as appropriate. Critical realism also emphasises the importance of contingency when making explanatory (causal) inferences (Sayer 1992); moreover, it does not prescribe on methods.

Giddens proposed structuration not as a replacement for other types of explanation in social theory, but as an account of what he was interested in: social structure. The 'problem' that this creates if one is interested in some degree of theoretical integration is that structuration theory is quite narrow in its possibilities, as mentioned above. Stones (2005) makes a suggestion that we follow up later in Chapter 4, namely, to use methodological bracketing to justify the juxtaposition of a sequence of studies that shed different types of light on a particular phenomenon, without prescribing the methods, disciplines, epistemologies or even ontologies that may underlie the separate studies. This bricoleur-type approach supports the same type of layered picture of reality as critical realism, leaving the MLP intact to function as it arguably does best, which is to summarily reflect and theorise – in a very general way – some of the processes involved in sociotechnical change. Of course, the MLP is limited, as is Giddens' (1984) account of social structure, and a critical realist ontology arguably does have the advantage of greater flexibility. Both strong (extended) structuration and critical realism, though, at least in our readings of the two, achieve the same end of supporting multiple types of explanation and multiple levels of analysis. It is these that are important when seeking to justify some degree of 'integration' of very different perspectives, in order to 'fill in the gaps' that systems-level sociotechnical transitions thinking leaves with insights from other disciplines.

One final question: Why fill in those gaps at all, in the context of energy or other sustainability transitions? Why not study sociotechnical phenomena in terms of the systems perspectives and high level frameworks that sociotechnical frameworks advocate, identifying the conditions under which different trends, processes or properties tend to emerge? Or, alternatively, why use sociotechnical perspectives at all? Why not simply frame analyses within the terms of other disciplines as appropriate? These are essentially specific instances of the question of the merits of interdisciplinarity: whether it is worth the difficulties and (at worst) superficialities that can accompany a lack of specialism and disciplinary knowledge.

No one is arguing here that disciplinary specialism is not useful. Nonetheless, the main disadvantage of disciplinarity without complementary interdisciplinarity is that phenomena important for explanation may be ignored. The same risk applies to explanations pitched at a general level. This is why, as Sorrell (2018) observes, researchers have been drawing attention to perspectives and issues that might enrich the generality of frameworks such as the MLP, to give more nuanced multi-level accounts. This combination of breadth and depth provides a higher level of understanding and a higher degree of confidence that the high level patterns posited reflect similar conditions and causal processes. This book extends this approach to a discipline that has rarely been brought together with sociotechnical frameworks to date, particularly in the context of energy – namely, social psychology. In the next sections, we provide an outline of the remainder of the book.

Outline of chapters

Part I: rationale

Chapter 2 draws on the large body of work on energy technology diffusion, acceptance and use and explains in more detail how these phenomena are currently approached. This literature is large and spans multiple contexts, methods, theoretical and disciplinary perspectives and paradigms. The purpose of the chapter is to provide an overview of this field as it is currently constituted; to explain the complementary but differing value of both psychological and sociological accounts; to discuss some of their methodological differences; and to lay the ground for the next chapter, which justifies, reviews and discusses the much smaller body of work that makes a case for informing sociotechnical transition studies with specifically psychological insights.

Chapter 3 begins by showing that in the sociotechnical sustainability transitions literature to date, social phenomena have mostly been examined in terms of relatively large collectives of people. This chapter documents and discusses the ways in which the psychology of agents or actors has been described and theorised within the sociotechnical transitions literature so far, both implicitly and explicitly. We show that while actor motivation and behaviour are often implicitly referred to, these are rarely theorised explicitly using psychological concepts. Although sociologists of science and technology have long understood technological diffusion and adoption as processes of social embedding, the psychological processes involved have received relatively little attention in the sociotechnical transitions literature. Reasons for the limited use of individual-level, psychological constructs are discussed.

Chapter 4 discusses existing frameworks that loosely connect differing perspectives, such as energy cultures and individual–social–material frameworks. Making our own contribution, we show how Stones' (2005) 'strong structuration' approach can be used to theorise the role of agency in sociotechnical perspectives of energy systems in a way that brings together psychological and sociological perspectives via methodological bracketing. In the sense in which we use it here, methodological bracketing denotes a sequence of closely related studies that may use very different perspectives and methods. Tentatively applying this approach, we begin to show how individual attitudes and beliefs in relation to niche energy technologies are influenced by national economic and innovation policy environments, so connecting the psychological and structural.

Part II: case study applications

Chapter 5 concerns the role of expectations in energy futures from a socio-technical perspective. 'Performative' here means 'helping to create': the literatures of innovation studies and science and technology studies suggest that positive expectations of new energy technologies will make it more likely that these will be developed, commercialised and used. In this chapter, we define expectations as beliefs about the future, and we show how the constructs of one of the most well-known psychological models of behaviour – the theory of planned behaviour – can help to explain why individuals act on some expectations, which then play a role in the mobilisation of resources for particular futures, while other expectations remain private and unrealised. We illustrate this with the case of stakeholder expectations of hydrogen fuel cell electric vehicles.

Chapter 6 applies Moscovici's (1988) theory of social representations as a social psychological concept that spans psychological and sociological levels, readily lending itself to connecting psychological and sociotechnical frames. We draw on energy examples from our own work and others, illustrating the roles of 'anchoring' and 'objectification' that are at the heart of the concept of social representations. We show how these concepts can help to explain not only the psychological aspects of energy controversies, but also how technological competition is associated with discursive competition, i.e. competing ways of interpreting and describing phenomena – in this case, energy technologies.

Chapter 7 takes another key psychological construct – values as cognitive–emotional phenomena – and explores their relevance for an exemplar grassroots innovation within the sharing economy. From a socio-technical perspective, grassroots innovations have been viewed as operating within 'niches'. Drawing on Martin and Upham (2016), we show why values matter in this context, and we discuss the extent to which values may or may not limit the scale up from the 'niche' to influencing the 'regime'. Our example uses the value scale of Schwartz (1992) and Freegle (Martin and Upham 2016) as a sharing economy activity that reduces energy consumption through product reuse, and we discuss the wider relevance for sustainability of what are regarded in sociotechnical thinking as slow-changing phenomena conceptually located in the background 'landscape' of taken-for-granted assumptions.

Chapter 8 discusses issues and methods of public engagement and participation at the research, development and deployment stages of new energy technology. Such issues have been the object of a gradually developing body of psychological work, and for several decades they have also been discussed in the science and technology studies literature. In sociotechnical transitions terms, transition management supports the engagement of a wide variety of stakeholders in policy development, as a necessary element for furthering sustainability. Using the example of lower carbon transport innovation policy, including new fuels, propulsion systems and associated taxes and incentives, this chapter discusses the connections between public opinion, public engagement and processes of creating transitions pathways. In particular, the chapter discusses how psychological surveying and opinion elicitation techniques might be adapted to provide information about sub-national differences in public opinion and public opinion heterogeneity for the purpose of informing energy and other policies.

Chapter 9 summarises the book and discusses possibilities, issues and directions for further research focusing particularly, by way of example, on social identity, values and their interconnections. We reiterate that we have sought to show that social psychology is not only relevant to sociotechnical transitions processes relating to energy and other sectors, but that the differences in analytic levels are bridgeable, in this case in the applied context of energy supply and use. From the premise that the behaviour change needed for successful energy transitions requires supportive action at all levels, we argue that the case for supportive, integrated and interdisciplinary analysis is strong. Finally, we also reiterate that we are very much aware that this book is only a first step, consolidating the beginnings of the research agenda that we propose, but with much further work to be done.

Notes

1 We use the term 'publics' to reflect the heterogeneity within and between those who are commonly thought of as 'the public'.
2 For example, by entrepreneurs, innovators and users; by agencies such as intermediaries; involving political contestation in specific contexts; or concerning the effects of specific regulation and policy on the progress of specific technologies.

Bibliography

Allport, G. W., Clark, K. and Pettigrew, T. 1954. *The Nature of Prejudice.* Reading, MA: Addison-Wesley.

Avelino, Flor and Rotmans, Jan. 2009. "Power in transition: An interdisciplinary framework to study power in relation to structural change". *European Journal of Social Theory* 12(4):543–69.

Bhaskar, R. 1975/1997. *A Realist Theory of Science.* London: Verso.

Bögel, Paula Maria and Upham, Paul. 2018. "Role of psychology in sociotechnical transitions literature: Review in relation to consumption and technology acceptance". *Environmental Innovation and Societal Transitions* 28:122–36.

Clayton, S., Devine-Wright, P., Stern, P. C., Whitmarsh, L., Carrico, A., Steg, L., Swim, J. and Bonnes, M. 2015. "Psychological research and global climate change". *Nature Climate Change* 5(7):640–6. doi:10.1038/nclimate2622.

Devine-Wright, P. 2009. "Beyond NIMBYism: Towards an integrated framework for understanding public perceptions of wind energy". *Wind Energy* 8:125–39. doi:10.1002/casp.1004.

Dosi, Giovanni. 1982. "Technological paradigms and technological trajectories. A suggested interpretation of the determinants and directions of technical change". *Research Policy* 11(3):147–62.

Figueres, C., Schellnhuber, H. J., Whiteman, G., Rockström, J., Hobley, A. and Rahmstorf, S. 2017. "Three years to safeguard our climate". *Nature* 546:593–5.

Frantzeskaki, Niki and Kabisch, Nadja. 2016. "Designing a knowledge co-production operating space for urban environmental governance – Lessons from Rotterdam, Netherlands and Berlin, Germany". *Environmental Science & Policy* 62:90–8. Available at: www.sciencedirect.com/science/article/pii/S1462901116300107.

Gazheli, Ardjan, Antal, Miklós and van den Bergh, Jeroen 2015. "The behavioral basis of policies fostering long-run transitions: Stakeholders, limited rationality and social context". *Futures* 69:14–30.

Geels, F. W. 2002. "Technological transitions as evolutionary reconfiguration processes: A multi-level perspective and a case-study". *Research Policy* 31(8–9):1257–74. doi:10.1016/S0048-7333(02)00062-8.

Geels, F. W. 2011. "The multi-level perspective on sustainability transitions: Responses to seven criticisms". *Environmental Innovation and Societal Transitions* 1(1):24–40. doi:10.1016/j.eist.2011.02.002.

Geels, F. W. 2014. "Regime resistance against low-carbon transitions: Introducing politics and power into the multi-level perspective". *Theory, Culture & Society* 31:21–40. doi:10.1177/0263276414531627.

Geels, F. W. and Schot, J. 2007. "Typology of sociotechnical transition pathways". *Research Policy* 36(3):399–417. doi:10.1016/j.respol.2007.01.003.

Giddens, A. 1984. *The Constitution of Society*. Cambridge: Polity Press.

Grin, J., Rotmans, J., Schot, J., Geels, F. and Loorbach, D. 2010. *Transitions to Sustainable Development: New Directions in the Study of Long Term Transformative Change*. New York: Routledge.

Heinrichs, Harald, Martens, Pim, Michelsen, Gerd and Wiek, Arnim (eds.) 2015. *Sustainability Science: An Introduction*. Dordrecht: Springer.

Hendriks, Carolyn M. 2008. "On inclusion and network governance: The democratic disconnect of Dutch energy transitions". *Public Administration* 86(4):1009–31. doi:10.1111/j.1467-9299.2008.00738.x.

Hynes, Mike. 2016. "Developing (tele)work? A multi-level sociotechnical perspective of telework in Ireland". *Research in Transportation Economics* 57:21–31.

Jacobsson, S. and Johnson, A. 2000. "The diffusion of renewable energy technology: An analytical framework and key issues for research". *Energy Policy* 28:625–40. doi:10.1016/S0301-4215(00)00041-0.

Kemp, René, Schot, Johan and Hoogma, Remco. 1998. "Regime shifts to sustainability through processes of niche formation: The approach of strategic niche management". *Technology Analysis & Strategic Management* 10:175–98.

Martin, C. J. and Upham, P. 2016. "Grassroots social innovation and the mobilisation of values in collaborative consumption: A conceptual model". *Journal of Cleaner Production* 134:204–13. doi:10.1016/j.jclepro.2015.04.062.

Moscovici, S. 1988. "Notes towards a description of social representations". *European Journal of Social Psychology* 18:211–50.

Nelson, Richard R. and Winter, Sidney G. 1982. *An Evolutionary Theory of Economic Change*. Cambridge, MA: Harvard University Press.

Nye, M., Whitmarsh, L. and Foxon, T. 2009. "Sociopsychological perspectives on the active roles of domestic actors in transition to a lower carbon electricity economy". *Environment and Planning A* 3:697–714. doi:10.1068/a4245.

Rockstrom, J., Steffen, W., Noone, K., Persson, A., Chapin, F. S. III, Lambin, E. F., Lenton, T. M., Scheffer, M., Folke, C., Schellnhuber, H. J., Nykvist, B., de Wit,

C. A., Hughes, T., van der Leeuw, S., Rodhe, H., Sörlin, S., Snyder, P. K., Costanza, R., Svedin, U., Falkenmark, M., Karlberg, L., Corell, R. W., Fabry, V. J., Hansen, J., Walker, B., Liverman, D., Richardson, K., Crutzen, P. and Foley, J. A. 2009. "A safe operating space for humanity". *Nature* 461:472–5.

Ross, L., Lepper, M. and Ward, A. 2010. "History of social psychology: Insights, challenges, and contributions to theory and application". In: S. T. Fiske, D. T. Gilbert and G. Lindzey (eds.), *Handbook of Social Psychology* (5th edn.). Hoboken, NJ: John Wiley & Sons, pp. 3–50.

Sayer, A. 1992. *Method in Social Science: A Realist Approach*. London: Routledge.

Schot, J. and Geels, F. W. 2008. "Strategic niche management and sustainable innovation journeys: Theory, findings, research agenda, and policy". *Technology Analysis & Strategic Management* 20(5):537–54. doi:10.1080/09537320802292651.

Schwartz, Shalom H. 1992. "Universals in the content and structure of values: Theoretical advances and empirical tests in 20 countries". *Advances in Experimental Social Psychology* 25:1–65.

Smith, Adrian, Stirling, Andy and Berkhout, Frans. 2005. "The governance of sustainable socio-technical transitions". *Research Policy* 34(10):1491–510. doi:10.1016/j.respol.2005.07.005.

Sorrell, S. 2018. Explaining sociotechnical transitions: A critical realist perspective. *Research Policy* 47(7):1267–82. doi:10.1016/j.respol.2018.04.008.

Sovacool, B. K. 2014. "What are we doing here? Analyzing fifteen years of energy scholarship and proposing a social science research agenda". *Energy Research & Social Science* 1:1–29. doi:10.1016/j.erss.2014.02.003.

Stangor, C. 2014. *Principles of Social Psychology – 1st International Edition* [ebook]. Pressbooks. Available at: https://opentextbc.ca/socialpsychology/.

Stedman, Richard C. 2016. "Subjectivity and social–ecological systems: A rigidity trap (and sense of place as a way out)". *Sustainability Science* 11(6):891–901.

Stephenson, J., Barton, B., Carrington, G., Doering, A., Ford, R., Hopkins, D., Lawson, R., McCarthy, A., Rees, D., Scott, M., Thorsnes, P., Walton, S., Williams, J. and Wooliscroft, B. 2015. "The energy cultures framework: Exploring the role of norms, practices and material culture in shaping energy behaviour in New Zealand". *Energy Research & Social Science* 7:117–23.

Stones, R. 2005. *Structuration Theory*. Basingstoke, UK: Palgrave Macmillan.

Svensson, O. and Nikoleris, A. 2018. "Structure reconsidered: Towards new foundations of explanatory transitions theory". *Research Policy* 47:462–73. doi:10.1016/j.respol.2017.12.007.

Toke, D. 2005. "Explaining wind power planning outcomes: Some findings from a study in England and Wales". *Energy Policy* 33:1527–39.

Ulli-beer, Silvia. 2013. "Conceptual grounds of socio-technical transitions and governance". In: Silvia Ulli-beer (ed.), *Dynamic Governance of Energy Technology Change*. Heidelberg: Springer, pp.19–47.

UNFCCC. 2015. "Paris Agreement". Conference of the Parties on its Twenty-First Session, Paris, 30 November to 11 December 2015. Available at: http://unfccc.int/resource/docs/2015/cop21/eng/l09r01.pdf.

United Nations General Assembly. 2015. "Transforming our world: The 2030 agenda for sustainable development". Available at: https://sustainabledevelopment.un.org/content/documents/21252030%20Agenda%20for%20Sustainable%20Development%20web.pdf.

Upham, P., Dütschke, E., Schneider, U., Oltra, C., Sala, R., Lores, M., Bögel, P. and Klapper, R. 2018. "Agency and structure in a sociotechnical transition: Hydrogen fuel cells, conjunctural knowledge and structuration in Europe". *Energy Research & Social Science* 37:163–74. doi:10.1016/j.erss.2017.09.040.

Van den Belt, H. and Rip, A. 1987. "The Nelson-Winter-Dosi model and the synthetic dye industry". In: W. E. Bijker, T. P. Hughes and T. Pinch (eds.), *The Social Construction of Technological Systems: New Directions in the Sociology and History of Technology*. Cambridge, MA: MIT Press, pp. 135–58.

Van Den Ende, J. and Kemp, R. 1999. "Technological transformations in history: How the computer regime grew out of existing computing regimes". *Research Policy* 28(8):833–51. doi:10.1016/S0048-7333(99)00027-X.

van der Horst, D. and Toke, D. 2010. "Exploring the landscape of wind farm developments: Local area characteristics and planning process outcomes in rural England". *Land Use Policy* 27:214–21. doi:10.1016/j.landusepol.2009.05.006.

Whitmarsh, Lorraine. 2012. "How useful is the multi-level perspective for transport and sustainability research?". *Journal of Transport Geography* 24:483–7. doi:10.1016/j.jtrangeo.2012.01.022.

2 Social science approaches to energy technology acceptance and diffusion

Introduction

The purpose of this chapter is to provide an overview of how psychology (mainly variable-based) is currently used in energy social science. This is principally in relation to energy technology acceptance and diffusion by different types of actors, most often publics. This provides context for the subsequent chapters, which illustrate connections between psychological and sociotechnical perspectives. The chapter draws on Upham *et al.* (2015) and our aim is again illustrative; it is not an attempt to review every psychological approach in the energy social science literature. We also explain the complementary but differing values of psychological and socio-logical accounts of human behaviour or practice, in order to make clearer the nature and implications of psychological-level analysis. In so doing, we discuss the difficulties of discussing energy technology acceptance in a theoretically neutral way.

Why do we need a social psychology of energy?

In answering this question, it would be easy to begin by saying that energy systems are changing, that people – and societies – need and consume energy, and that they accept *and* support processes of transition to sustain-able energy resources. It would also be good to say that the world is heading towards a low carbon or even renewables-based future. Yet while there are real grounds for optimism in this regard (normatively speaking), the fact is that the largest percentage change in primary energy supply at a global level over the last 40 years has been a shift from fossil oil towards fossil natural gas (IEA 2017). There are also some worrying trends at the time of writing (mid-2018) (Vaughan 2018). Nonetheless, the energy mix in many countries is changing, even if belatedly, towards renewables. Globally, between 2005 and 2015 wind electricity production in terawatt hours (TWh) grew eight-fold and global solar photovoltaic (PV) electricity production grew by a staggering 62 times (IEA 2017). So energy systems are changing, albeit too slowly for climate targets.

In the context of energy system change, a large body of psychological and indeed sociological work has examined public acceptance of – and objection to – energy technologies in various contexts, from country level to the local and household level (Lorenzoni and Pidgeon 2006; Devine-Wright 2013). The number of relevant studies is substantial, and they span public attitudes to and levels of acceptance of nuclear energy, hydrogen, carbon capture and storage (CCS), wind, biomass plants and other renewable and low carbon energy technologies (Upham *et al.* 2009; Whitmarsh *et al.* 2011). Similarly, a wide variety of studies based on different approaches and methodologies, mainly case studies, has addressed key elements in the interaction between energy developments and host communities (Walker and Devine-Wright 2008). While the more sociological approaches to studies of such themes have generally not used or emphasised the term 'acceptance', the insights gained from these studies have a strong bearing on the understanding of energy technology and related policy acceptance – including critiques of the psychological perspective that gives rise to the term 'acceptance' (e.g. Shove *et al.* 2008).

Nonetheless, the recognition that public acceptance does play a role in technology development, installation and use raises many questions about the complexity of the processes that shape public responses to energy technologies and infrastructures at different levels (Devine-Wright 2009). It has also raised questions about the policy and practical implications (Batel *et al.* 2013), as well as questions about the conceptual, definitional and methodological basis of research on social and public acceptance in this area (Wüstenhagen *et al.* 2007; Devine-Wright 2008; Hitzeroth and Megerle 2013).

In this chapter we draw heavily on Wüstenhagen *et al.* (2007) to set out a simple, applied framework that is primarily rooted in social psychology and that explains something of the complementarity of psychological and sociological perspectives, while at the same time recognising that at specific levels of attributed causality and conception, sociological, psychological and technical accounts have marked and ultimately irreconcilable differences (Shove 2010; Shove and Walker 2014). Accordingly, we emphasise, distinguish and classify 'acceptance' in terms of the principal levels at which acceptance can be studied. Furthermore, we distinguish the main classes of the referential object – distinctions that are apparently simple, but which we nonetheless consider too often obscured by variable-level and case-specific detail. For this purpose, we provide an overview of theories and perspectives that we believe to be influential. As stated above, the overall purpose of the chapter is to provide a knowledge base on the social psychology of energy, plus some contrast with complementary sociological perspectives, such that the challenges to theoretical integration with sociotechnical thinking are clearer.

Technology acceptance

The idea of technology 'acceptance' is itself not theoretically neutral, but has its origins in broader ideas of technology diffusion and the associated empirical investigation of variables such as the perceived usefulness of the technologies (Davis 1989). Nonetheless, the term has become widespread and so we take it as a starting point in our discussion of theory (see below), drawing on an illustrative literature review by Upham *et al.* (2015). We concur with the pragmatic principle of the bricoleur of acknowledging value in a range of perspectives (Rogers 2012). In other words, we see the value of multiple perspectives and methods as shedding different types of light on different aspects of a problem.

A wide variety of disciplinary and indeed paradigmatic perspectives have been applied to understanding individual and societal reactions to energy technologies and developments (Lutzenhiser 1993; Wilson and Dowlatabadi 2007). These include microeconomics and behavioural economics (emphasising rational choice models, investment behaviour and pricing policy); environmental sociology (emphasising equity, process, policy and institutions as well as practice and habit-related theory and social norms); and social psychology (emphasising motivation, risk perception, place and identity and behavioural theories connecting norms, values, behaviour and a variety of mediating variables).[1] Below we provide an overview of this variety of acceptance-related perspectives, as a precursor to setting out a general framework of technology acceptance. The primary focus is on energy technology acceptance. Table 2.1 summarises some notable contributions, with the selection reflecting Upham *et al.* (2015), Upham *et al.* (2009) and Whitmarsh *et al.* (2011).

The first modern references to technology acceptance were in relation to utility-focused contexts of technology implementation and adoption (Davis 1989), followed by perspectives relating to the diffusion of innovation (Rogers 1995). This simplistic, but quite influential, linear view sees acceptance as occurring at the second and third stages of the sequence: (1) knowledge; (2) persuasion; (3) decision stages; (4) implementation; and (5) subsequent confirmation (Rogers 1995). The heuristic also allows us to locate much of the psychological and behavioural economics literature at the *persuasion* and *decision* stages. By contrast, the sociological literature gives greater emphasis to the ways in which individuals are connected to – and are a product of – their social and physical (including technological) environments.

As suggested above, many of the differences among these perspectives cannot be completely bridged while retaining their own terms. Most notably from a behavioural psychological perspective, particularly in variants of one of the most commonly used theories, the theory of planned behaviour (Azjen 2005), a key focus is understanding the relationships between psychological constructs – notably attitudes, norms, beliefs and

Table 2.1 An illustrative selection of perspectives on acceptance of energy technologies

Discipline	Perspective and illustrative authors	Synopsis
Economics	Choice models (e.g. Labay and Kinnear 1981).	• Individuals form preferences regarding energy technologies by making trade-offs between the various attributes of those technologies. • Consumers are expected to act based on logically determined and articulated preferences of utility.
	Behavioural economics (e.g. Frederiks *et al.* 2015).	• Modifies the above assumptions of economic rationality to account for psychological factors.
Environmental sociology and human geography	Equity, process, policy and institutions (Walker 2007); practice and habit as part of social structuration (Shove and Southerton 2000; Warde 2005; Shove 2010); socio-demographics and lifestyles (Claudy *et al.* 2010); environmental conflict and land use planning systems (Wolsink 2000; Upham and Shackley 2006; Haggett 2008; van der Horst and Toke 2010 etc.).	A wide-ranging set of perspectives that include attention to: • the social, economic, political and technological context of individuals that shape and constrain attitudes and behavioural responses to low carbon energy and associated risks; • 'practices' approaches from the sociology of consumption, in which behaviour, habits and routines are viewed as shaping attitudes, rather than vice versa; • participatory engagement, structures of ownership, the distribution of benefits and other institutional factors; • various types of social influence processes, including social norms; • socio-demographic characteristics such as age, gender and social class; • lifestyles, habits and needs; • resistance as a function of local, contextual factors.

continued

Table 2.1 Continued

Discipline	Perspective and illustrative authors	Synopsis
Social psychology	Theories of planned behavior and norm activation (de Groot and Steg 2008); risk perception (Pidgeon et al. 2008); environmental concern, values, norms, behaviour (Stern 2000); place identity and attachment (Devine-Wright 2013); social representations (Castro 2006; Batel et al. 2013).	A wide range of models and perspectives, focusing, for example, on: • attitude, social and personal norms, perceived behavioural control and intention; • personal, emotional attachments to places and their role in individual identity; • subjective judgments of the characteristics and severity of technological risk.
Cultural theory	Application of Mary Douglas' cultural theory approach (West et al. 2010).	• Cultural worldviews as attitudinal determinants.
Other frameworks and methods driven work	The eclectic energy cultures approach (Stephenson et al. 2010); communications theory and information processing (Brunsting et al. 2011); use of Q-sort to characterise positions (Cuppen et al. 2010); use of informed choice questionnaires (de Best-Waldhober et al. 2009 etc.).	• Many studies, often in the grey literature, take no explicit theoretical stance, although attitude theory is usually implicit. The examples listed here are more conscious of theory but either seek to avoid strong mono-theoretical subscription or are heavily methods driven.

Source: Upham *et al.* 2015.

Note

The relevant literature amounts to thousands of papers; these are a personal, illustrative selection. Substantial reviews are available that also consider, tangentially, relevant perspectives such as science and technology studies (Upham *et al.* 2009; Whitmarsh *et al.* 2011).

values – as a precursor to intention and behaviour. Yet from the perspective of the sociology of practice, behaviour is viewed in terms of the practices that structure society (Bourdieu 1977; Shove and Walker 2014). In response, one way of making a social psychological account of energy technology acceptance somewhat more neutral is to frame its presentation in terms that emphasise the *contexts* of acceptance. This is what we do below (Upham *et al.* 2015).

Basic elements of an analytical framework

Why use the term 'acceptance' in the context of energy technologies if it cannot adequately capture the range of research perspectives? Principally, because the term has been used so widely, and in so many different contexts, that it is difficult to replace. Despite the variety of work on social acceptance in relation to new energy technologies, the concept of acceptance itself has usually been taken for granted (Ricci *et al.* 2008; Batel *et al.* 2013). Researchers and stakeholders alike use the concept of acceptance to refer to a range of objects: in relation to lay public attitudes towards energy technologies, either in the abstract or implicitly or explicitly in relation to policy support; in relation to the position of policy actors on investments in specific energy technologies; in relation to support for and opposition to specific energy developments at the local level; and/or in relation to the diffusion of energy applications at the household or the organisational level. Indeed, the concept of acceptance often seems to be treated as a process disconnected from other dimensions of social response to energy technologies.

With the above in mind, Upham *et al.* (2015) suggest the following definition of *acceptance*:

> a favourable or positive response (including attitude, intention, behavior and – where appropriate – use) relating to a proposed or in situ technology or socio-technical system, by members of a given social unit (country or region, community or town and household, organization).

At its simplest, acceptance thus defined may simply denote a lack of active opposition, although it may also take more active forms of expression such as political support. Importantly, acceptance takes place in a particular social and political context.

Although acceptance of energy technologies has become a focus of attention, partly for instrumental reasons – energy technology advocates and developers benefit from or require public acceptance of their technologies – acceptance of energy technologies is just one part of the broader phenomenon of how individuals, groups and societies interact with energy developments (Williams and Edge 1996; Pursell 1999). Indeed, acceptance involves multi-dimensional, dynamic processes that are obscured by the use of a simple phrase. As Batel *et al.* (2013) argue:

> if we keep focusing on this term [social acceptance] – either purposefully or not – we are not only perpetuating the normative top-down perspective on people's relations with energy infrastructures, but we are also potentially ignoring all the other types of responses to those, such as support, or uncertainty, resistance, apathy, among others.

Yet, despite its simplicity, 'acceptance' is so prevalent a term that it is difficult to dispense with.

Using the operational definition of acceptance above, three general principles relating to the social acceptance of energy technologies can be established:

1   *The social acceptance of a technology can be analysed at three levels: macro, meso and micro*, typically corresponding to: (a) the general, policy or country, level; (b) the community, town or other geographically defined level; and (c) the individual entity level, such as households or organisations. These levels tend to correlate with different objects of acceptance: respectively, types of energy supply technology; specific energy infrastructure proposals or installations; and on-site energy applications that may be demand or supply side.

2   *Social acceptance at the three levels may refer to the following differentiated components* – depending on the subject of the acceptance: (a) public acceptance, in the sense of individual consumers and citizens; (b) stakeholder acceptance, in the sense of organisations without formal political objectives but with an interest in the outcome; and (c) political acceptance, in the sense of policy support by governmental levels, agencies and political parties.

3   *The internal structure of individual acceptance is composed of attitudinal elements (attitudinal acceptance), behavioural intentions and actual behaviours (behavioural acceptance).* Acceptance at this level includes beliefs and feelings (cognition and affect) about and in relation to an energy supply technology, infrastructure development or application, but also the willingness to accept or use the technology, and actual (public-sphere and private-sphere) behaviour (Upham *et al.* 2015).

It is clear from the above that there are different levels (or units of analysis) at which acceptance of energy technologies can be analysed. For example, a country may reject nuclear energy as a matter of policy, or a local community may oppose a specific shale gas project or homeowners may or may not install small-scale wind energy applications. All these processes refer to the social acceptance of specific technologies, but the different levels at which social acceptance is referred to involve different processes and components of social acceptance. Previous efforts to differentiate the various levels include Devine-Wright (2007), Wüstenhagen *et al.* (2007), Seyfang *et al.* (2007), Wolsink (2007) and Stern (2014). Reviewing these and other work, we concur with previous proposals that there are three *typical levels* of acceptance analysis, particularly that of Wüstenhagen *et al.* (2007), developing these as follows:

1   Acceptance of an energy supply technology at the policy or national level (*social acceptance*). At this level, acceptance research has typically sought to understand the levels of social acceptance (including the general public, policy makers, civil society organisations, experts, private organisations, etc.) at the country, state or regional level towards a particular energy supply technology. The technology is typically considered in general and in aggregate. For example, a particular country may or may not accept (invest, support, etc.) nuclear energy or offshore wind. Individuals and representatives in this country may perceive that the technology may, or may not, be acceptable at a general level.

2   Acceptance of an energy infrastructure or facility at the local level (*community acceptance*). At this level, acceptance research has sought to understand the reaction of communities (comprising local decision makers, local stakeholders and local citizens) towards a particular proposed energy infrastructure. Research questions are related to the reaction of a community (city, small town, etc.) towards a specific energy infrastructure. For example, the reaction of a community towards a wind development, a proposed CO_2 storage site, a shale gas extraction project, etc. The focus here is on the interaction of a community (including the individuals and the stakeholders that shape it) with physical fuel extraction, supply, production, conversion or storage infrastructure, or a project proposal in relation to these.

3   Acceptance of an energy application at the household and organisation level (*market acceptance*). Research at this level has sought to investigate the reaction of actual and potential end-users and stakeholders (such as householders, investors or plant managers) towards particular demand and supply side energy applications (e.g. micro-generation technologies or more efficient appliances). The object of acceptance here is typically a specific energy application that can be installed within a home, business or organisation and to which utility criteria are applied (Upham *et al.* 2015).

Figure 2.1 summarises the above, focusing on the *contexts* of acceptance following Wüstenhagen *et al.* (2007), while Table 2.1 (Upham *et al.* 2015) better reflects those research fields of energy technology acceptance that are active and distinct (Gaede and Rowlands 2018).

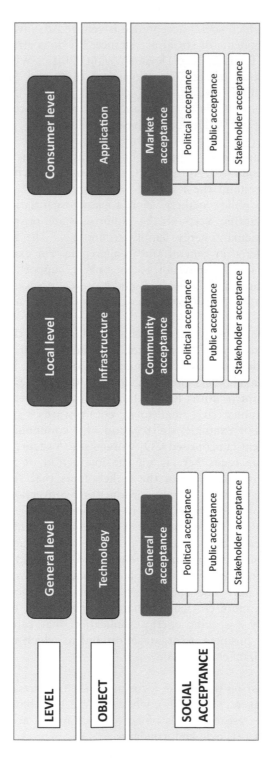

Figure 2.1 A context-based classification of types of energy technology acceptance.

Source: Upham *et al.* 2015, following Wüstenhagen *et al.* 2007.

The subjects of social acceptance – those who are 'doing the accepting' – thus span a wide-ranging set of groups: individual members of the public; professional end-users; the many types of civil society groups (including community groups, non-governmental organisations (NGOs), labour unions, indigenous groups, charitable organisations, faith-based organisations, professional associations and foundations); and companies and industry associations, politicians, academia, etc. These groups may constitute categories of people with particular, common characteristics, or they may consist of organised groups. They can be further, more generally categorised in terms of: *publics*; *organised political groups*; and *other organised stakeholder groups* (commercial, non-commercial and mixed). Following from this, we can distinguish between types of social acceptance: public acceptance, political acceptance and other stakeholder acceptance. This classification is a heuristic, in that an actor may fall into more than one group given the social role that they play at a particular time. Thus, the actors in Table 2.2, which provides a summary mapping of actor groups and social acceptance, may belong to more than one group.

In this classification, regulatory and political organisations and institutions operating at different scales, from the local to the national or international, provide a first component of societal acceptance – *political acceptance*. This refers to the attitude or behavioural response towards the implementation or adoption of a proposed technology by decision makers and key members of the political system in a given society, community or

Table 2.2 Actor groups and social acceptance at the three levels

Levels/actors	General/policy level	Local/community level	Household/organisation/end-user level
Political actors	*National acceptance* by national, formally instituted decision makers.	*Local political acceptance* by local, formally instituted decision makers.	*User acceptance* by individual citizens with views on energy policy.
Stakeholders	*Stakeholder acceptance* by other nationally active market and non-market policy groups.	*Local stakeholder acceptance* by other locally active market and non-market policy groups.	*Stakeholder acceptance* by commercial and other organised users.
Publics	*Public acceptance* by the general population as citizens with views on national policy.	*Local public acceptance* by the local population as citizens with views on national policy.	*End-user acceptance* by household and individual end-users.

town. *Stakeholder acceptance* refers to the members of the stakeholder groups in a social unit, that is, in a particular country or town. This might include the various groups of civil society, companies and industry associations that can affect – or be affected by – the proposed technology or development. Finally, *public acceptance* refers to the attitude or behavioural response to the implementation or adoption of a proposed technology by the lay public of a given country, region or town. Individuals act as citizens who react in different ways to energy policies, technologies and infrastructures developed in their countries or cities (Stern 2014; Upham *et al.* 2015).

At the household and organisation level, acceptance by the end-users, including professional-users and lay-users, is the key component of social acceptance. Some end-users also play a more active role, modifying or adapting technologies (Hyysalo *et al.* 2013). Besides end-user acceptance, stakeholder acceptance also plays a role at this level, referring to acceptance by investors, industry, local government and other affected actors.

Challenges to theoretical integration

The section above has set out an analytical framework with which to both study the social acceptance of energy technologies, infrastructures and applications, and with which to classify studies of these. The framework draws directly on Wüstenhagen *et al.* (2007) but also reflects experience of the many empirical and conceptual studies undertaken since. The framework deliberately simplifies and generalises and hence obscures a number of methodological and analytical challenges, which we discuss below. The following sections review some of these challenges and we also propose recommendations for future research directions and for practical resolution. The research challenges listed are not intended to be exhaustive in terms of social and behavioural research in relation to energy – broader lists are available elsewhere (e.g. Sovacool 2014) that emphasise issues that are important in the context of public (lay public or end-user) acceptance of energy technologies.

The psychology of energy transitions, addressed from a sociotechnical perspective, is inevitably cross-disciplinary and also of applied interest. Researchers in this field are generally interested in informing both practice and policy and they want their work and their perspectives to be widely understood. This raises a number of issues for knowledge users from outside the social sciences who seek to make sense of what has now become a large, applied field of research. Indeed, a prioritisation of this sense-making informs the decision (Upham *et al.* 2015) to build on the conceptualisation of energy technology acceptance proposed by Wüstenhagen *et al.* (2007) and retain use of the term 'acceptance', despite its limitations.

At the risk of over-simplifying and also entering philosophy of science debates that we do not want to engage with in depth here, some social scientists seek to follow the norms and methods of the natural sciences;

others, particularly more constructivist and/or qualitative perspectives within the social sciences, do not seek to make the same claims of generalisability and repeatability and are more interested in providing in-depth accounts than in identifying regularities that apply over multiple cases, albeit perhaps only under particular conditions. Moreover, the social sciences seek to contribute to knowledge not only through empirical novelty, but also through conceptual novelty. In the social sciences, then, new ways of thinking, seeing the world through novel and eye-opening analytical perspectives, are also viewed as useful. Accordingly, the degree of conceptual, methodological and explanatory diversity in the social sciences is high, and for those seeking purely unified or integrated insights into the same phenomena, this is an undesirable situation (Gintis 2007).

In this book, we are looking at forms of conceptual or theoretical integration that allow contributory research and analytical perspectives to retain their integrity, by which we mean that the terms of those perspectives are respected. A relevant example in the context of energy is the ISM – Individual Social Material – approach (Southerton *et al.* 2011). This is a loose framework that gives equal weight to insights from different disciplines when seeking to understand environmentally relevant behaviour. Similarly, the early version of the energy cultures framework (Stephenson *et al.* 2010) does not (arguably) attempt a high level of integrative theorisation, but rather is intended to bring a variety of factors together. These examples notwithstanding, integrative work – and especially close integrative work – is rather rare in this field and, as a result, challenges remain, especially for knowledge users but also for researchers. We discuss some of these below.

Box 2.1 Energy cultures framework

With a cultural focus, "[t]he 'Energy Cultures' conceptual framework aims to assist in understanding the factors that influence energy consumption behaviour and to help identify opportunities for behaviour change" (Stephenson *et al.* 2010; see also Stephenson *et al.* 2015; Stephenson 2018). In a broad sense, the multidisciplinary energy cultures framework (ECF) addresses the question: Why do people become 'locked into' particular patterns of energy use and consumption?

The EFC aims to facilitate an understanding of why people sometimes continue with their old habits of energy consumption/energy behaviours, even when change seems logical to an observer. To this end, the ECF considers the relevant aspects of energy-related *social norms* (e.g. what is normal and appropriate), *material culture* (e.g. technologies and infrastructure) and *practices* (i.e. what people do when using energy). Here,

> practices are understood as everyday actions (both routinized and less frequent) that are common across social peers. It [the ECF] also incorporates the acquisition of the material objects that enable people to

> enact and reproduce those practices. The material culture are those available objects, technologies ... that can control, influence, and affect people's energy demand. The cognitive norms are expectations of a particular service or behaviours that are shaped by a specific meaning attributed to them. As such, within the ECF, cognitive norms include values, beliefs and attitudes.... A particular set of these [norms, material cultures and energy practices combined] with the external influences, give rise to a distinct energy culture.
>
> (Stephenson *et al.* 2010; see also Ambrosio-Albala *et al.* 2019)
>
> In this way, the ECF allows for the study of energy consumers as autonomous agents within a wider social, cultural, political and material/ technological context, but with the emphatic recognition that all of these factors influence behaviours associated with energy consumption. In other words, a particular combination of these 'softer' and 'harder' factors identified within the ECF may either push for change in energy-related behaviours, and thus for an energy culture change, or rather they may support energy culture stasis or resistance. In short, the ECF facilitates an understanding of why people use energy the way that they do, with an implied objective that this understanding is intended to pave the way for energy behaviour change.
>
> (Stephenson *et al.* 2010; Hopkins and Stephenson 2014)

Diversity in social science energy research

Research on public attitudes towards energy technologies has relied on a wide variety of research designs and methods. Examples include experimental and quasi-experimental designs in relation to carbon capture and storage (CCS) (L'Orange Seigo *et al.* 2013) and nuclear energy (Showers and Shrigley 1995); observational and correlational designs based on conventional surveys (Bollinger and Gillingham 2012); information choice questionnaires (de Best-Waldhober *et al.* 2009) and Q-method (Venables *et al.* 2009); case-specific questionnaires (Upham and Shackley 2006); mixed method designs based on qualitative case studies and quantitative field studies (Upham and Shackley 2007); and combining questionnaires and interviews of end-users (Mourato *et al.* 2004) and others.

Research diversity is not a challenge or problem per se. On the contrary, research diversity may indicate a very healthy research field. However, drawing consistent interpretations across studies from results obtained via such diverse methods and perspectives does raise a number of questions and challenges regarding complementarity and immediate research integration. Each design and method produces a particular type of knowledge, framed in a particular way, with different purposes, scopes, limitations and conditionality (Mahoney 2008). Below we make a number of observations in this regard, by way of illustration, focusing in particular on some of the issues raised when seeking to integrate data obtained via different methods:

- Particularly for knowledge users outside of – or new to – the social sciences, the most challenging problem arises when seeking to combine insights from qualitative and quantitative research, which each have (often very) different claims to research generalisability.
- Compounding this is the way in which critical, typically qualitative researchers may challenge quantitative claims to generalisability (see Symonds and Gorard 2010). As Gorard (2006) argues: "research involving numbers is as interpretivist and about meaning and judgement as much as research without numbers". Quantitative social science may be interpreted in more than one way and (as mentioned) is likely to have been produced under particular conditions that may or not pertain (or be believed to pertain) in other situations.
- While large-scale surveys offer statistical representativeness and the potential for inferential models, they are usually by necessity based on a relatively shallow level of questioning and are highly conditional on specific question phrasing. Modifying phrasing or the information presented with the questions may well give different results.
- In quantitative social science, numerical scales readily omit certain important aspects of phenomena. For example, most scaled questions cannot capture the intricate details of individual experience and emotions that may underpin respondent answers to some questions.
- In qualitative research, the researcher may simply miss something of importance for the phenomena studied, or perhaps the researcher cannot negotiate access to key information or social fora. Inevitably, through the representation of qualitative research insights, and perhaps through coding practices, the essence of much qualitative research is condensed, and much qualitative information lost.
- When done well, however, qualitative research will capture and cast light on more qualitative phenomena present in and important for society, and quantitative research will demonstrate very real societal tendencies that can be apprehended in a more numerical manner. What matters most for both qualitative and quantitative research are the decisions and judgements involved in careful research design and framing, the identification of relevant empirical themes and the subsequent analytical choices.
- Case studies allow in-depth examination of process and situational factors, but making generalisable inferences needs to be done tentatively (Sartori 1994), and this is perhaps best done at an abstract or attribute level, with general relevance primarily being with respect to similar cases (Yin 2009). Case-oriented researchers are often interested in the causal processes involved in particular outcomes in specific cases, such that they need to be sensitive to time, place, agency and process (Allardt 1990; Ragin 1999). They can nonetheless build upon multiple sources and types of data and use mixed methods.

- Mixed methods studies seek to combine research depth and breadth, and as such they do require additional resources. The strength of such studies is that they ideally utilise the merits of both quantitative and qualitative research approaches and, accordingly, compensate for their respective weaknesses too. However, mixed methods research has tended to raise objections or concerns in research communities that are traditionally either quantitatively or qualitative inclined (Johnson and Onwuegbuzie 2004).

- Importantly, qualitative, quantitative and mixed methods research approaches all have methodological strengths and weaknesses, and the specific object of study – and the research question – should determine what approach is most suitable for any particular research endeavour (Johnson and Onwuegbuzie 2004). In short, little should be taken for granted when seeking to integrate results derived from different research methods, and critical reflection by the knowledge users as to the applicability of the chosen research methods for the particular phenomena studied is always beneficial.

Theoretical diversity compounds issues of commensurability. Studies often focus on different aspects of a research problem, but whether their results, conclusions and interpretation should be treated as alternatives or complementary in a quasi-summative way is sometimes debatable. For example, consider, on the one hand, studies of public responses to CCS that emphasise the role of information and rational choice (de Best-Waldhober *et al.* 2009) and, on the other hand, studies that emphasise trust (Terwel *et al.* 2009). In this example, we might conclude that both trust and information are important in the context of CCS (which they undoubtedly are), but this does not do full justice to either study. So, if extensive, guided provision of information does lead respondents to reluctantly accept CCS, though not in preference to renewables (de Best-Waldhober *et al.* 2009), should we conclude that a lack of trust can be overcome if sufficient information is provided? To do so would be to contradict the widespread critique of the (information) deficit model in relation to behaviour change (Kollmuss and Agyeman 2002). Yet it is also likely that information does play an important role in this and related contexts, particularly where scientific and technical knowledge matter (Sturgis and Allum 2004). In all likelihood, the roles of information and trust are nuanced, conditional, interactive, dynamic and variable. It is in these and other interactions that there is potential for further research, particularly regarding the *conditions* under which different factors have different degrees of influence on the attitudes, beliefs and/or behaviour of technology acceptance. Add to this other theoretical perspectives, even within social psychology, such as place attachment with its own correlates (Devine-Wright 2013), and one can see that even a single example raises questions about knowledge integration – and also the limits to this (Hoyningen-Huene 2002; Upham *et al.* 2015).

Summary

In this chapter, we have set out a simple, social psychological conceptual account of understanding different types of acceptance of new energy technologies. In so doing, we have provided both an overview of the field that we want to bring into sociotechnical transitions thinking and an introduction to some of the challenges in doing so.

Despite wide use of the term 'acceptance' in relation to energy project and technology developments within and outside academia, the definition and meaning of the term has often been taken for granted. The acceptance framework distinguishes between three levels of social acceptance (general, local and household/organisational/end-user levels), based on a distinction between the acceptance of an energy technology, an infrastructure or an application. It also makes a distinction between three components of social acceptance at each level based on the referent of acceptance (political, stakeholder and public acceptance) and, finally, it tackles the characterisation of the internal structure of individual acceptance. The framework uses the main *contexts* of acceptance as categories, to minimise theoretical subscription and, hence, maximise applicability across perspectives.

Our motivation here is that the social sciences have tended to frame energy technology acceptance studies from particular perspectives (Wilson and Dowlatabadi 2007) rather than engaging in an interdisciplinary, problem-oriented effort to develop an integrated understanding of the social acceptance of energy technologies (Stern 2014; Gaede and Rowlands 2018). No *single* analytical approach will provide a satisfactory framework for analysing more than a fraction of individual and social phenomena. Nor can a single analytical approach fully underpin successful policy interventions (Biggart and Lutzenhiser 2007; Wilson and Dowlatabadi 2007; Stephenson *et al.* 2010). While it is unrealistic and undesirable to have complete unanimity of theoretical perspectives in this context, there is a role for conceptual frameworks that integrate the many factors that are influential for social and public acceptance of emerging energy technologies (Devine-Wright 2007). Such integrative approaches should be able to span categories of acceptance of energy technologies and consider the multitude of influences on the attitudes and behaviours involved. This breadth is in turn a precondition for understanding the interacting effects of issues of interest to most environmental social science researchers, regardless of discipline (Stern 2014; Upham *et al.* 2015).

In the next chapter we move on from the social psychology of energy technology acceptance as the field is now, to look at what use sociotechnical sustainability transitions theorists are making of it. Given the importance of energy systems for sustainability, one might imagine that this use would be considerable. Of course, as we highlighted above, psychological perspectives are little used in a transitions context, and we both document this and consider why.

Note

1 A comparative, summary account of policy and project-based interventions drawing on some of these perspectives can be found at www.ieadsm.org/ViewTask.aspx?ID=17&Task=24&Sort=0.

Bibliography

Allardt, E. 1990. "Challenges for comparative social research". *Acta Sociologica* 33:183–93.

Ambrosio-Albala, P., Upham, P. and Bale, C. 2019. "Purely ornamental? Public perceptions of distributed energy storage in the United Kingdom". *Energy Research & Social Science* 48:139–50.

Azjen, I. 2005. *Attitudes, Personality and Behaviour.* Maidenhead, UK: Open University Press.

Batel, Susana, Devine-Wright, Patrick and Tangeland, Torvald. 2013. "Social acceptance of low carbon energy and associated infrastructures: A critical discussion". *Energy Policy* 58:1–5.

Biggart, N. W. and Lutzenhiser, L. 2007. "Economic sociology and the social problem of energy inefficiency". *American Behavioral Scientist* 50:1070–87.

Bollinger, Bryan and Gillingham, Kenneth. 2012. "Peer effects in the diffusion of solar photovoltaic panels". *Marketing Science* 31(6):900–12. doi:10.1287/mksc.1120.0727.

Bourdieu, P. 1977. *Outline of a Theory of Practice.* Cambridge: Cambridge University Press.

Brunsting, S., De Best Waldhober, M., Upham, P., Dütschke, E., Oltra, C., Desbarats, J., Riesch, H. and Reiner, D. 2011. "Communicating CCS: Applying communications theory to public perceptions of carbon capture and storage". *International Journal of Greenhouse Gas Control* 5:1651–62. doi:10.1016/j.ijggc.2011.09.012.

Castro, P. 2006. "Applying social psychology to the study of environmental concern and environmental worldviews: Contributions from the social representations approach". *Journal of Community and Applied Social Psychology* 16:247–66.

Claudy, M. C., Michelsen, C., Driscoll, A. O. and Mullen, M. R. 2010. "Consumer awareness in the adoption of microgeneration technologies: An empirical investigation in the Republic of Ireland". *Renewable and Sustainable Energy Reviews* 14:2154–60.

Cuppen, E., Breukers, S., Hisschemöller, M. and Bergsma, E. 2010. "Q methodology to select participants for a stakeholder dialogue on energy options from biomass in the Netherlands". *Ecological Economics* 69:579–91. doi:10.1016/j.ecolecon.2009.09.005.

Davis, Fred D. 1989. "Perceived usefulness, perceived ease of use, and user acceptance". *MIS Quarterly* 13(3):319–39.

de Best-Waldhober, Marjolein, Daamen, Dancker and Faaij, André. 2009. "Informed and uninformed public opinions on CO_2 capture and storage technologies in the Netherlands". *International Journal of Greenhouse Gas Control* 3(3):322–32. doi:10.1016/j.ijggc.2008.09.001.

de Groot, J. I. M. and Steg, L. 2008. "Value orientations to explain beliefs related to environmental significant behavior: How to measure egoistic, altruistic, and biospheric value orientations". *Environment and Behavior* 40:330–54. doi:10.1177/0013916506297831.

Devine-Wright, Patrick. 2007. "Reconsidering public attitudes and public acceptance of renewable energy technologies : A critical review". A working paper of the research project "Beyond Nimbyism: A multidisciplinary investigation of public engagement with renewable energy technologies" funded by the Economic & Social Research Council. Available at: http://geography.exeter.ac.uk/beyond_nimbyism/deliverables/bn_wp1_4.pdf.

Devine-Wright, Patrick. 2008. "Reconsidering public acceptance of renewable energy technologies: A critical review". In: M. Grubb, T. Jamasb and M. G. Pollitt (eds.), *Delivering a Low Carbon Electricity System: Technologies, Economics and Policy*. Cambridge: Cambridge University Press.

Devine-Wright, Patrick. 2009. "Beyond NIMBYism: Towards an integrated framework for understanding public perceptions of wind energy". *Wind Energy* 8(2):125–39. doi:10.1002/we.124.

Devine-Wright, Patrick. 2013. "Explaining 'NIMBY' objections to a power line: The role of personal, place attachment and project-related factors". *Environment and Behavior* 45(6):761–81. doi:10.1177/0013916512440435.

Eisenhardt, Kathleen M. and Graebner, Melissa E. 2007. "Theory building from cases: Opportunities and challenges". *Academy of Management Journal* 50(1):25–32. doi:10.5465/amj.2007.24160888.

Falconer, D. J. and Mackay, D. R. 1999. "The key to the mixed method dilemma". In: Proceedings of the 10th Australasian Conference on Information Systems, 1–3 December 1999, School of Communications and Information Management, Te Kura Whakaipurangi Kōrero, Victoria University of Wellington, New Zealand, pp. 286–97.

Frederiks, E. R., Stenner, K. and Hobman, E. V. 2015. "Household energy use: Applying behavioural economics to understand consumer decision-making and behaviour". *Renewable and Sustainable Energy Reviews* 41:1385–94. doi:10.1016/j.rser.2014.09.026.

Gaede, James and Rowlands, Ian H. 2018. "Visualizing social acceptance research: A bibliometric review of the social acceptance literature for energy technology and fuels". *Energy Research and Social Science* 40:142–58.

Gintis, H. 2007. "A framework for the unification of the behavioral sciences". *Behavioral and Brain Sciences* 30:1–16. doi:10.1017/S0140525X07000581.

Gorard, Stephen. 2006. "Towards a judgement-based statistical analysis". *British Journal of Sociology of Education* 27:67–80.

Haggett, C. 2008. "Over the sea and far away? A consideration of the planning, politics and public perception of offshore wind farms". *Journal of Environmental Policy & Planning* 10:289–306. doi:10.1080/15239080802242787.

Hitzeroth, Marion and Megerle, Andreas. 2013. "Renewable energy projects: Acceptance risks and their management". *Renewable and Sustainable Energy Reviews* 27:576–84.

Hopkins, D. and Stephenson, J. 2014. "Generation Y mobilities through the lens of energy cultures: A preliminary exploration of mobility cultures". *Journal of Transport Geography* 38:88–91. doi:10.1016/j.jtrangeo.2014.05.013.

Hoyningen-Huene, Paul. 2002. "Paul Feyerabend und Thomas Kuhn". *Journal for General Philosophy of Science/Zeitschrift für allgemeine Wissenschaftstheorie* 33:61–83.

Hyysalo, S., Juntunen, J. K. and Freeman, S. 2013 "User innovation in sustainable home energy technologies". *Energy Policy* 55:490–500. doi:10.1016/j.enpol. 2012.12.038.

IEA. 2017. "World energy outlook". International Energy Agency, Paris.

Jacobsson, Staffan and Johnson, Anna 2000. "The diffusion of renewable energy technology: An analytical framework and key issues for research". *Energy Policy* 28:625–40.

Johnson, R. Burke and Onwuegbuzie, Anthony J. 2004. "Mixed methods research: A research paradigm whose time has come". *Educational Researcher* 33:14–26. doi:10.3102/0013189X033007014.

Kollmuss, A. and Agyeman, J. 2002. "Mind the gap: Why do people act environmentally and what are the barriers to pro-environmental behavior?". *Environmental Education Research* 8(3):239–60.

Labay, D. G. and Kinnear, T. C. 1981. "Exploring the consumer decision process in the adoption of solar energy systems". *Journal of Consumer Research* 8:271.

L'Orange Seigo, Selma, Dohle, Simone, Diamond, Larryn and Siegrist, Michael. 2013. "The effect of figures in CCS communication". *International Journal of Greenhouse Gas Control* 16:83–90.

Lorenzoni, I. and Pidgeon, N. 2006. "Public views on climate change: European and USA perspectives". *Climatic Change* 77(1–2):73–95.

Lutzenhiser, L. 1993. "Social and behavioural aspects of energy use". *Annual Review of Energy and the Environment* 18:247–89.

Mahoney, J. 2008. "Toward a unified theory of causality". *Comparative Political Studies* 41:412–36.

Mourato, Susana, Saynor, Bob and Hart, David. 2004. "Greening London's black cabs: A study of driver's preferences for fuel cell taxis". *Energy Policy* 32:685–95.

Pidgeon, N. F., Lorenzoni, I. and Poortinga, W. 2008. "Climate change or nuclear power – No thanks! A quantitative study of public perceptions and risk framing in Britain". *Global Environmental Change* 18:69–85. doi:10.1016/j. gloenvcha.2007.09.005.

Pursell, Carroll W. 1999. "Beyond engineering: How society shapes technology". *Technology and Culture* 40:395–7.

Ragin, Charles C. 1999. "The distinctiveness of case-oriented research". *Health Services Research* 34:1137–51.

Ricci, Miriam, Bellaby, Paul and Flynn, Rob. 2008. "What do we know about public perceptions and acceptance of hydrogen? A critical review and new case study evidence". *International Journal of Hydrogen Energy* 33(21):5868–80.

Rogers, E. M. 1995. *Diffusion of Innovations*. New York: Free Press.

Rogers, Matt. 2012. "Contextualizing theories and practices of bricolage research". *The Qualitative Report* 17:1–17.

Sartori, G. 1994. "Comparison and comparative method". In: G. Sartori and L. Morlino (eds.), *La Comparación En Las Ciencias Sociales*. Madrid: Alianza, pp. 29–47.

Seyfang, G., Lorenzoni, I. and Nye. M. 2007. "Personal carbon trading: Notional concept or workable proposition? Exploring theoretical, ideological and practical underpinnings". Centre for Social and Economic Research on the Global Environment Working Paper EDM 07–03, University of East Anglia, Norwich, UK.

Shove, E. 2010. "Beyond the ABC: Climate change policy and theories of social change". *Environment and Planning A* 42:1273–85.

Shove, E. and Southerton, D. 2000. "Defrosting the freezer: From novelty to convenience – A narrative of normalization". *Journal of Material Culture* 5:301–19.

Shove, E. and Walker, G. 2014. "What is energy for? Social practice and energy demand". *Theory, Culture & Society* 31:41–58.

Shove, E., Chappells, H., Lutzenhiser, L. and Hackett, B. 2008. "Comfort in a lower carbon society". *Building Research & Information* 36(4):307–11.

Showers, Dennis E. and Shrigley, Robert L. 1995. "Effects of knowledge and persuasion on high-school students' attitudes toward nuclear power plants". *Journal of Research in Science Teaching* 32:29–43. doi:10.1002/tea.3660320105.

Southerton, D., McMeekin, A. and Evans, D. 2011. "International review of behaviour change initiatives". The Scottish Government, Edinburgh.

Sovacool, Benjamin K. 2014. "What are we doing here? Analyzing fifteen years of energy scholarship and proposing a social science research agenda". *Energy Research & Social Science* 1:1–29. doi:10.1016/j.erss.2014.02.003.

Stephenson, J. 2018. "Sustainability cultures and energy research: An actor-centred interpretation of cultural theory". *Energy Research & Social Science* 44:242–9. doi:10.1016/j.erss.2018.05.034.

Stephenson, J., Barton, B., Carrington, G., Gnoth, D., Lawson, R. and Thorsnes, P. 2010. "Energy cultures: A framework for understanding energy behaviours". *Energy Policy* 38:6120–9. doi:10.1016/j.enpol.2010.05.069.

Stephenson, J., Barton, B., Carrington, G., Doering, A., Ford, R., Hopkins, D., Lawson, R., McCarthy, A., Rees, D., Scott, M., Thorsnes, P., Walton, S., Williams, J. and Wooliscroft, B. 2015. "The energy cultures framework: Exploring the role of norms, practices and material culture in shaping energy behaviour in New Zealand". *Energy Research & Social Science* 7:117–23.

Stern, P. 2000. "Toward a coherent theory of environmentally significant behaviour". *Journal of Social Issues* 56:407–24.

Stern, Paul C. 2014. "Individual and household interactions with energy systems: Toward integrated understanding". *Energy Research & Social Science* 1:41–8.

Sturgis, Patrick and Allum, Nick. 2004. "Science in society: Re-evaluating the deficit model of public attitudes". *Public Understanding of Science* 13(1):55–74. doi:10.1177/0963662504042690.

Symonds, Jennifer E. and Gorard, Stephen. 2010. "Death of mixed methods? Or the rebirth of research as a craft". *Evaluation & Research in Education* 23(2):121–36. doi:10.1080/09500790.2010.483514.

Terwel, B. W., Harinck, F., Ellemers N. and Daamen, D. D. L. 2009. "Competence-based and integrity-based trust as predictors of acceptance of carbon dioxide capture and storage (CCS)". *Risk Analysis* 29(8):1129–40. doi:10.1111/j.1539-6924.2009.01256.x.

Toke, Dave. 2005. "Explaining wind power planning outcomes: Some findings from a study in England and Wales". *Energy Policy* 33(12):1527–39.

Upham, P. and Shackley, S. 2006. "Stakeholder opinion on a proposed 21.5 MWe biomass gasifier in Winkleigh, Devon: Implications for bioenergy planning and policy". *Journal of Environmental Policy and Planning* 8(1):45–66.

Upham, P. and Shackley, S. 2007. "Local public opinion of a proposed 21.5 MW(e) biomass gasifier in Devon: Questionnaire survey results". *Biomass & Bioenergy* 31(6):433–41.

Upham, P., Whitmarsh, L., Poortinga, W., Purdam, K. and Devine-Wright, P. 2009. "Public attitudes to environmental change: A selective review of theory and practice". A research synthesis for the Living with Environmental Change Programme. Available at: https://esrc.ukri.org/files/public-engagement/public-dialogues/full-report-public-attitudes-to-environmental-change/.

Upham, P., Oltra, C. and Boso, À. 2015. "Towards a cross-paradigmatic framework of the social acceptance of energy systems". *Energy Research & Social Science* 8:100–12. doi:10.1016/j.erss.2015.05.003.

van der Horst, Dan and Toke, David. 2010. "Exploring the landscape of wind farm developments: Local area characteristics and planning process outcomes in rural England". *Land Use Policy* 27:214–21.

Vaughan, A. 2018. "IEA warns of 'worrying trend' as global investment in renewables falls", *Guardian* [online], 17 July 2018. Available at: www.theguardian.com/business/2018/jul/17/iea-warns-of-worrying-trend-as-global-investment-in-renewables-falls.

Venables, Dan, Pidgeon, Nick, Simmons, Peter, Henwood, Karen and Parkhill, Karen. 2009. "Living with nuclear power: A q-method study of local community perceptions". *Risk Analysis* 29:1089–104.

Warde, A. 2005. "Consumption and theories of practice". *Journal of Consumer Culture* 5(2):131–53. doi:10.1177/1469540505053090.

Walker, G. P. 2007. "Environmental justice and the distributional deficit in policy appraisal in the UK". *Environmental Research Letters* 2:045004.

Walker, Gordon and Devine-Wright, Patrick. 2008. "Community renewable energy: What should it mean?". *Energy Policy* 36(2):497–500.

West, J., Bailey, I. and Winter, M. 2010. "Renewable energy policy and public perceptions of renewable energy: A cultural theory approach". *Energy Policy* 38:5739–48.

Whitmarsh, L., Upham, P., Poortinga, W., McLachlan, C., Darnton, A., Devine-Wright, P., Demski, C. and Sherry-Brennan, F. 2011. "Public attitudes to and engagement with low-carbon energy: A selective review of academic and non-academic literatures". Report for RCUK Energy Programme, London. Available at: https://orca.cf.ac.uk/22753/1/EnergySynthesisFINAL20110124.pdf.

Williams, Robin and Edge, David. 1996. "The social shaping of technology". *Research Policy* 25:865–99.

Wilson, Charlie and Dowlatabadi, Hadi. 2007. "Models of decision making and residential energy use". *Annual Review of Environment and Resources* 32:169–203.

Wolsink, M. 2007. "Wind power implementation: The nature of public attitudes: Equity and fairness instead of 'backyard motives'". *Renewable and Sustainable Energy Reviews* 11:1188–207.

Wolsink, M. 2000. "Wind power and the NIMBY-myth: Institutional capacity and the limited significance of public support". *Renewable Energy* 21:49–64.

Wüstenhagen, Rolf, Wolsink, Maarten and Bürer, Mary Jean. 2007. "Social acceptance of renewable energy innovation: An introduction to the concept". *Energy Policy* 35:2683–91.

Yin, Robert K. 2009. *Case Study Research: Design and Methods*. Thousand Oaks, CA: Sage Publications.

3 How is social psychology currently used in the sociotechnical sustainability transitions literature?

Introduction

The *social* is by definition inherent to – and a key focus of – sociotechnical perspectives, and a key premise of sociotechnical transitions thinking is that society and technology co-evolve. The idea of co-evolution builds upon observations of the more specific interaction of social and technological design choices, as studied in the social construction of technology (SCOT) tradition (Pinch and Bijker 1984; Bijker *et al.* 1987). More broadly, sociotechnical transitions thinking holds that technological developments lead to social and broader societal responses, which then affect the further development of technology and so on (Adil and Ko 2016). This duality is reflected in theoretical frameworks such as the multi-level perspective (MLP) (Geels 2002), where (as discussed in Chapter 1) it is argued that transitions come about through different types of interaction between processes at three levels: niche-protected innovations gradually becoming more powerful; landscape-level change that puts pressure on the sociotechnical regime; and/or destabilisation of the regime enabling niche-innovations to gain their own momentum (Rip *et al.* 1995; Bögel and Upham 2018). The processes involved in these changes are not only viewed as both social and technological, but as involving interrelationships between the two.

While sociological accounts engage with processes experienced by individuals, the origins and nature of those accounts are (by definition) posited as social in the sense of involving more than one person. Sociotechnical transition researchers do acknowledge the role of subjective human experience, but from sociological perspectives. For example, the roles of meanings, interpretation, discourses and symbols are referred to (Stedman 2016), but these processes are understood from social perspectives. Yet human experience in transitions processes can also be examined from psychological (particularly social psychological) perspectives (Gazheli *et al.* 2015). There are of course deep ontological differences between psychological and sociological perspectives (for example, regarding the nature of agency; for a more detailed discussion on ontological differences and also points of connection, see e.g. Rivers 1916; Thoits 1995; Shove 2010).

With this in mind, this chapter examines the ways in which social psychology has been used within the sociotechnical sustainability transitions field to date. It also examines the reasons why that use has been limited. Our premise – again – is that psychological explanations of various aspects of individual agency have their own intrinsic value, that while there may be sociological analogues or equivalents for understanding particular sociotechnical processes, "ignoring insights from psychological research can handicap progress towards a low-carbon, sustainable future" (Clayton *et al.* 2015). While sociological or cultural accounts of subjectively experienced phenomena place their focus external to the individual in terms of processes and emphases, psychology emphasises the characteristics and processes of individuals (micro-level) or groups of individuals (meso-level): "sociologists generally devote their efforts to identifying *which* social phenomena have effects on individuals while psychologists generally specialize in identifying *the mechanisms or processes through which* social phenomena have their effect on individuals" (Thoits 1995, p. 1231, cited in Mangone 2017, p. 23).

In this chapter, we show that psychological theory and empirics have been little used in the sociotechnical transitions literature, and we suggest that instead the emphasis in this literature has been on various forms of collective agency (see Smith *et al.* 2005; Hynes 2016). Collective agency expressed through institutions and organisations lends itself to explanatory accounts that involve shared, social processes. In addition, the systems view of sociotechnical research contrasts with the tendency of (much, though not all) psychology to focus on the isolated effects of single factors. This in turn follows from the field's dominant methodological approach, namely, forms of experiments (including exploratory attitudinal surveys) that aim at causality testing.

Indeed, a justifiable critique of the psychological paradigm that is probably most well known in sustainability-related psychology is that its individualistic and cognitive nature (Keller *et al.* 2016) fails to adequately: (a) take the influence of the environment into account (Sorrell 2015); and (b) consider the influence that individuals can have on that environment (co-construction or, in our context, co-evolution). While concurring with this critique, we nonetheless take the view that the internal dynamics of subjective experience do matter for sociotechnical accounts and processes; therefore, there is merit in considering how such processes may be better and more closely integrated into structural and collective accounts, including through the use of a broader range of psychological theories in the sociotechnical transitions literature than those considered so far.

The chapter examines how the small number of studies referring to psychological perspectives in the sociotechnical transitions literature use a rather limited range of psychological theories, and that these are appealed to mostly in relation to a functional perspective (i.e. in terms of outcomes). The studies do not yet examine the psychological processes themselves in depth (Nye *et al.* 2010; Whitmarsh 2012; Gazheli *et al.* 2015; Stephenson *et al.* 2015). More especially, they do not yet seek to connect those psychological processes to sociotechnical processes in any depth.

To this end, we conduct a systematic literature review to document and then discuss the ways in which the psychology of agents or actors is described and theorised in the sociotechnical transitions literature, both implicitly and explicitly. We show that the primary use of psychology in the sociotechnical transitions literature has been in relation to consumption and technology acceptance. Of the large variety of perspectives and theories available to psychology, only six (or so) main theoretical perspectives have been deployed in the sociotechnical transitions literature, of which one – social practice theory – is sociological in origin, albeit arguably with a psychological component relating to habitus and dispositions (Bourdieu 1986). In the latter part of the chapter, we show how more use could be made of these six particular approaches, though we leave a fuller discussion of broader research directions for the final chapter. The chapter as a whole draws on Bögel and Upham (2018).

Method

First, a note on methods. To identify the use of psychology in the sociotechnical transitions literature we examined a relevant literature database (Scopus) for the co-concurrence of relevant keywords. As is always the case with such studies, the choice of keywords, database and further interpretative decisions all affect one's conclusions, and thus the conclusions only hold under the conditions of the particular choices made. For example, the chosen source database, Scopus, does not reflect the arts, humanities or the social sciences as comprehensively as Google Scholar's database. Perhaps more importantly, there may be a large body of papers focusing on aspects of sociotechnical transitions that do not use the term 'sociotechnical' explicitly: that is, they are relevant to the subject, but framed differently. This characteristic will apply to much of the environmental psychology and sustainable consumption psychology literatures. It will also apply to literatures using, for example, behavioural economic perspectives that may omit the term 'psych' from the key search fields. The search conducted here is thus quite narrowly focused, partly for pragmatic reasons but also because we have an explicitly specific intention.

We searched the Scopus database in the spring of 2017 using the search terms 'sociotechnical transitions' and 'psych*' as an indicator of explicit use of psychological theory in the transitions literature. First, we applied these search terms to the 'abstract, title and keywords' field of the search facility in Scopus. This led to only two papers, lending some support to our starting hypothesis that psychological theories are rarely explicitly referred to in sociotechnical transitions research. To capture additional documents with an *implicit* use of psychological theories, the search string was then applied to 'all fields'. In order to refine the research topic, the search was limited to journal articles. The publication time period was left open, and thus limited only by the parameters of the database. However, the sociotechnical transitions literature in its current form appeared only in the early 2000s, despite the term 'sociotechnical' having been coined much earlier (Emery and Trist 1960) in the context of coal-mining.

While the above search method omits papers that could be considered relevant, it nonetheless returned 191 results in the defined terms. Most publications appeared in the journals *Energy Policy* (12) and *Technological Forecasting and Social Change* (12), followed by the *Journal of Cleaner Production* (10) and also *Energy Research & Social Science* (8), despite the latter having been founded only relatively recently. Overall, the literature review shows an emerging body of relevant research over the last 10 years, with a maximum of nearly 50 publications in 2016 (see Figures 3.1 and 3.2). Most publications originated in the UK (74), followed by the United States (28), the Netherlands (20), Australia (18) and Germany (18). We have not undertaken any bibliometric research to examine co-citation patterns and whether more specific sub-fields might be identified.

Having found the above papers, we first categorised studies of limited relevance to the topic ($n = 87$). A second category consisted of studies that use psychological theories implicitly ($n = 81$): this is comprised of studies that refer to constructs addressed in the psychological literature, e.g. attitudes or values, but where specific psychological theories and/or findings

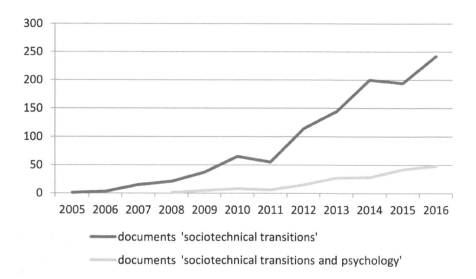

Figure 3.1 Numerical count of articles in the Scopus database.

Note
The search used the keywords 'sociotechnical transitions' (keywords only) and 'sociotechnical transitions + psych*' (all fields); search conducted 8 March 2017.

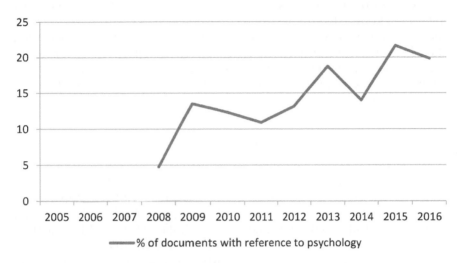

% of documents with reference to psychology

Figure 3.2 Numerical count of articles in the Scopus database (represented as annual percentages).

Note
The search used the keywords 'sociotechnical transitions' (keywords only) and 'sociotechnical transitions + psych*' (all fields); search conducted 8 March 2017.

are mostly not referenced, or where psychological approaches are applied, albeit only with a brief reference. The fact that most of these manuscripts do implicitly use psychological theories supports our starting hypothesis that the sociotechnical transitions literature oftentimes refers implicitly to psychological constructs, acknowledging their importance, but does not use them substantively. The third category of studies explicitly used psychological theories within a sociotechnical transitions frame ($n = 23$); even here, though, the use of psychological theories was light. We discuss categories two and three further below; our distinctions between the categories are not definitive, but are to help characterise the situation.

Implicit use of psychological constructs and theories

Light implicit acknowledgement

Within this category, two different ways of using psychological insights can be found. Here, attitudes and behaviours are referred to as being highly relevant to transitions research, but neither psychological theories nor empirics are connected to the main scientific work. For example, in their analysis of the further development of battery-electric vehicles, Nilsson and Nykvist (2016) discuss barriers and drivers of the diffusion of battery-electric vehicles and – among others – several psychologically relevant constructs such as knowledge, norms and cognitive barriers. The authors acknowledge the importance of consumer-citizen knowledge for the diffusion of battery-electric vehicles and refer to ways of increasing this. However, neither relevant psychological empirics nor theories are used to develop or inform inferences, with the purpose of the authors' study being to demonstrate the value of the MLP in scenario development (Nilsson and Nykvist 2016).

More detailed implicit acknowledgement

In the second category, psychological constructs are referred to in a little more depth, but still without *explicit* reference to psychological theory or empirics. The types of psychological constructs referred to in this category include: motives (e.g. Chatterton 2016; Ruppert-Winkel *et al.* 2016), values (e.g. Audet and Guyonnaud 2013; Miller *et al.* 2014; Lilliestam *et al.* 2016), trust (e.g. Büscher and Sumpf 2015; Goedkoop and Devine-Wright 2016; Missimer *et al.* 2017), attitudes (Hassink *et al.* 2013; Ma *et al.* 2013; Bowerman 2014; Puhe and Schippl 2014; Scott *et al.* 2014) and well-being (Dawson and Martin 2015; Brown and Vergragt 2016; Folke 2016; Macmillan *et al.* 2016; Bögel and Upham 2018).

Below we comment on the studies in this second category, presented according to their theoretical foundations. We consider in turn for their potential for integration in sociotechnical transitions research. To be clear, this selection does not exhaust the possibilities of psychological approaches, but it does reflect those approaches most used to date in socio-technical transitions research (Bögel and Upham 2018).

1 RATIONAL AND MINDFUL DECISION MAKING

Studies in this category rely on the characterization of people as *homo oeconomiucs*, defined by rational, deliberate and mindful decisions and actions. Most often, these studies use the theory of planned behaviour (TpB) by Ajzen (1991, 2011) as a central theoretical approach, probably the most popular psychological theory for behaviour change in rational, conscious situations. Meta studies also suggest that it is the most successful approach in terms of statistically predicting behaviour (Webb *et al.* 2010). The TpB relies on the assumption that how one evaluates the consequences of a potential change in one's behaviour will strongly influence behavioural intentions and actual behaviour. The approach is based on the expectancy-value approach (Atkinson 1957) that broadly posits the same assumption. The TpB more specifically assumes that human behaviour is guided by three factors: attitude towards the behaviour (beliefs about outcomes × outcome likelihood); social norms (beliefs about what others think); and perceived behavioural control (beliefs about how easy or difficult it will be to carry out the behaviour in question).

Studies using TpB focus on individual behaviour, often consumption behaviour. The TpB has been one of the most influential approaches in the early stages of sustainable consumption research (Liu *et al.* 2017), and it is still used in this regard in sociotechnical transitions research, e.g. renewable energy system adoption by consumers. For example, Yun and Lee (2015) coupled the sociotechnical perspective with the TpB to analyse consumers' adoption of renewable energy systems. The authors confirm the influence of attitudes, subjective norms and perceived behavioural control on consumers' intentions to use renewable energy systems (Yun and Lee 2015). In addition, they identify variables that have an impact on attitude (social trust, social support) and behavioural control (facilitating technical conditions). These latter, additional findings are particularly relevant for guiding interventions that aim at increasing behavioural intention by changing one of the three factors cited above (Bögel and Upham 2018).

While the focus of the TpB is usually at the micro (individual) level, Brones *et al.* (2017) use of the theory to examine changes at the organisational level. In their study, the authors analyse sustainability change, namely, the integration of eco-design, at an organizational level, from a transitions perspective. They suggest that individual behaviour change theories should be combined with organisational change theories for this purpose. Brones *et al.* (2017) develop an eco-design transition framework that includes these different theoretical approaches, and they apply it to a case study. Their framework is a good example of how theories from transitions research can be combined with individual behaviour change approaches (Bögel and Upham 2018).

Use of the TpB in the transitions literature notwithstanding, there are challenges. Most notably, such stark individual-level theories do not easily connect with the collective and aggregated units of analysis – notably organisations, institutions and system processes – of sociotechnical transitions perspectives. There are also specific aspects of the TpB that are problematic. First, the fundamental assumption of the TpB is the rational and conscious decision making of individuals. Yet such decision making, requiring awareness and close attention, cannot always be assumed (Gazheli *et al.* 2015; Sorrell 2015). Consumption behaviour, for example, is often habitual (Audet and Guyonnaud 2013). Second, although the social norm constructs in the current version of the theory aid, in principle, connection to the sociotechnical (e.g. Sopha *et al.* 2013), the social processes involved in this are rarely considered in depth (Jackson 2005; Axsen and Kurani 2014). Seeking to remedy this, Axsen and Kurani (2014) propose three different layers and types of intrapersonal influence: awareness of a new technology, as dependent on societal diffusion; assessment, involving negotiation and interpretation of meaning of the new technology; and, finally, a change in self-concept via a process of reflexivity.

2 HABITUAL BEHAVIOUR

The extent to which pro-environmental behaviour can be understood as either deliberate or habitual behaviour has gained attention in sociotechnical transitions research, and energy research in particular. Hence, Nye *et al.* (2010) suggest distinguishing between different forms of energy behaviour: deliberate energy *conservation* behaviour and ongoing energy *use* behaviour in everyday life. To the extent that energy use behaviour is habitual (see also Sorrell 2015), theories that assume conscious, reflective decision making, such as the TpB, may not offer reliable forms of explanation.

Instead, Nye *et al.* (2010) suggest that a new stream of psychological research that focuses on the social construction of identity and consumption may offer a more promising way to research habitual behaviour. The underlying premise here is that identity and lifestyle aspects are key drivers of energy behaviour. Hence, Nye *et al.* (2010) refer to the example of air conditioning as a symbol of modern life and lighting as a symbol of prosperity in addition to their utilitarian functions. Similarly, Koutiva *et al.* (2017) show the influence of identity processes in water use, be this either conscious or habitual.

Another perspective of relevance to habitual behaviour is social practice theory. Although sociological in its origins and intentions, its prime advocate, Bourdieu (1986), sought to explain social structure through social practice theory. But here it is the *individuals* that are considered in relation to practices and to their social inheritance (their 'habitus'). Social practice theory has been widely used in sustainability research to explain habitual behaviour (Kent 2015). In social practice theory, "routine human action is understood as a product of collective social practices influenced as much by the environment as it is by personal preferences or processes of deliberation" (Kent 2015, p. 729). Practices are thus a key unit of analysis. However, this also underpins one of the main critiques of Bourdieusian practice theory, namely, that individual agency in the sense of self-determination is conceived of as highly constrained: changes come from 'above' (Batel *et al.* 2016).[1]

In her study on automobility, Kent (2015) attempts to overcome this limitation by combining social practice theory with psychological theories. The author examines how feelings such as those of comfort, defined as bodily sensations, are an essential part of the practice of private car use. More generally, this would seem a promising approach – combining social practice theory with psychological approaches that "point to the importance of identity, lifestyles, and 'subjective norms' in driving everyday, 'unthinking' energy use" (Nye *et al.* 2010, p. 702; see Bögel and Upham 2018).

3 THE ROLE OF NORMS

In contrast to the expectancy-value approach, norm-based approaches assume that action is not motivated by tangible or social outcomes, but instead by more internal drivers associated with personal norms. Note that personal norms differ from social norms as defined in the TpB. While social norms are characterised as beliefs about what others think one should do, personal norms are an expression of one's internal values and may also reflect the internalisation of social norms. Some integrative frameworks, such as the energy cultures framework (ECF), also mix personal and social norms. With reference to the ECF, Stephenson *et al.* (2015, p. 357) provide an example: "Norms are personal and social expectations about how life should be lived, including (in relation to transport) such things as expectations about speed of travel and aspirations for car ownership."

There are several psychological theories that focus on the normative dimension of pro-environmental behaviour, such as Ecological Value Theory, Schwartz's Norm Activation Theory and Stern's Value Belief Norm Theory (for an overview, see Jackson 2005). The latter explicitly refers to the influence of values on personal norms. While theories relating to norms and values have been widely used in research on pro-environmental behaviour they are, however, less present in sociotechnical transitions research. As mentioned above, values are often implicitly referred to, but very few studies within sociotechnical transitions frames explicitly address value theories from psychological perspectives. An exception is Martin and Upham (2016), who discuss the roles of citizens' and activists' values, and the differences between these, in the diffusion of grassroots social innovations. The authors examine a product reuse initiative in this light.

4 THE SOCIETAL LEVEL: SOCIAL REPRESENTATIONS THEORY

While the TpB is an individual-level psychological theory, social representations theory (Moscovici 1988) is a societal level social psychological theory. Put simply, social representations can be described as "complex ideas, processes and objects translated into common sense that is accessible and applicable in everyday life" (Evensen and Stedman 2016, p. 15). Two main processes are relevant in social representations theory: *anchoring*, the linkage of a new concept, such as shale gas (Upham *et al.* 2015), to existing concepts that thereby lend familiarity to novel ideas, and *objectification*, the use of imagery and/or descriptive language to summarily represent an idea. Both processes occur as part of social communication and discourse.

Several authors have suggested that social representations theory may be combined with individual-level psychological theories. For example, social representations theory has been discussed as an extension to cultural perspectives (Sarrica *et al.* 2016), such as the ECF (Stephenson *et al.* 2010). The ECF is an interdisciplinary approach that is intended to bridge individual and societal levels (Stephenson *et al.* 2015). The interaction between norms (e.g. shared beliefs, habits), material culture (availability of technologies relevant to energy use, e.g. public transport system) and energy practices is reflected in the ECF.

Moreover, social representations theory offers a less implicitly passive and contextually focused view of individuals than social practice theory. As Sarrica *et al.* (2016, p. 3) comment regarding social practice theory:

> a further step is needed to achieve a more effective comprehension of the individual–society link; the cultural factors should not been seen and studied only as external data embedded in the organization of material and social life that influences individual feelings, judgment and behaviours but also a constitutive part of the individual mind.

In social representations theory, while representations are defined as culturally and contextually situated, the theory also accounts for transformations based on communication processes and negotiations between individuals and groups, conceptually leaving room for individual agents to act (Bögel and Upham 2018).

Another example of the combination of social representations theory with individual psychological theories is the work of Evensen and Stedman (2016). These authors combine social representations theory with construal level theory to examine public perceptions of shale gas development. Construal level theory implies that objects physically closer to a person or nearer in time are (unsurprisingly) perceived in more detail and specificity, and hence considered more salient. Likewise, socially relevant self-experienced activities or phenomena are described and thought about in more detail than socially distant experiences. Construal theory has, for example, been referenced as part of an explanation of why climate change has been perceived as spatially and temporally remote (Whitmarsh and Lorenzoni 2010). Evensen and Stedman (2016) use both construal level theory and social representations theory to show that close proximity strengthens the relationship between perceptions of shale gas development and support for/opposition to development. This should not be seen as offering support for a simplistic 'NIMBY' thesis of public objection, but illustrates the bringing together of individual and social level analyses as a means of understanding public responses to energy technology.

Other socio-psychological approaches have also been applied to study the role of citizen and consumer expectations and beliefs in sociotechnical transitions research. Phillips and Dickie (2015) compare the narratives of people with different levels of action aimed at carbon reduction, e.g. the use of public transport, or building or moving to an eco-house, in order to understand attitude–behaviour gaps. The authors found that people use different narratives to deal with the dissonance aroused by differences between environmental concerns and action.

5 PLACE ATTACHMENT

The concept of place plays an important role in sociotechnical transitions (see e.g. Murphy and Smith 2013; Feola and Nunes 2014; Whitmarsh *et al.* 2015; Süsser *et al.* 2017). Outside of a sociotechnical transitions frame, Devine-Wright (e.g. 2009) has conducted an extended programme of place-related research in relation to public perceptions of new energy infrastructure. Place identity and attachment approaches in sociotechnical transitions research deal with the question of how people/peoples are attached to places, and how this influences publics' perceptions of actual and proposed changes to those places. For example, in their study on shale gas hydraulic fracking, Whitmarsh *et al.* (2015) find that place-based factors (location, rurality, employment in energy industry, length of residence in area and place attachment) strongly influence social acceptance of shale gas. Similarly, Feola and Nunes (2014) show the influence of place on the success of grassroots social innovations; for example, that less successful or completely inactive transitions initiatives are more likely to be in urban areas. The authors assume that the reason for this is partially due to weaker local attachments to place in urban areas. Likewise, a study by Süsser *et al.* (2017) confirms the influence of meaning of place on the emergence of grassroots innovations in the context of renewable energies (Bögel and Upham 2018).

Place attachment research approaches often combine geographical and psychological factors (Devine-Wright 2009; Whitmarsh *et al.* 2015), and it is this interdisciplinary perspective that has probably favoured their application in sociotechnical transitions research. Devine-Wright (2009) points out how place approaches and social representations theory could be used together to study place attachment from societal-level perspectives, particularly vis-à-vis processes of interpretation, evaluation and contestation. This argument is essentially the same as that used by Evensen and Stedman (2016) in the context of public perceptions of fracking for shale gas. Place-based approaches have been suggested as a way to foster agency-related research on sociotechnical systems by other researchers too (see Stedman 2016). Indeed, if agency at the individual level comes to be accounted for more often in sociotechnical transitions research, then the role of place and related context is a key candidate as a future research direction (Bögel and Upham 2018).

6 INFORMATION, PERSUASION AND COGNITIVE DISSONANCE

Several studies undertaken within sociotechnical transitions frames have found a gap between (high) environmental awareness and concern and levels of action to mitigate climate change, and have referred to dissonance processes in this regard (see Phillips and Dickie 2015). Cognitive dissonance theory, initially developed by Leon Festinger (1957), proposes that if a person holds two cognitions that are related but inconsistent, s/he experiences cognitive dissonance. As cognitive dissonance is an unpleasant state, people consciously or unconsciously seek to reduce this by changing some aspect of their cognition. According to Festinger's original definition, "A cognition can be any bit of knowledge people may hold about the world, their more immediate environment, or themselves, including their attitudes, beliefs, behaviours, and affective states" (Festinger 1957, cited in Nail and Boniecki 2011, p. 46; see Bögel *et al.* 2018).

While social psychologists have studied cognitive dissonance as an individual phenomenon, Cohen (2013, p. 43) suggests further examination of dissonance at the societal level, so called 'collective dissonance'. Cohen identifies collective dissonances as one factor that will lead to change processes within sociotechnical systems, because "pronounced levels of collective dissonance cannot be maintained indefinitely" (Cohen 2013, p. 43). One might say that Cohen's proposition is thus a psychological expression of the pressures and tensions on and within the sociotechnical regime, such pressures notably arising from the landscape or niche levels.

A small number of studies that apply a sociotechnical transitions research framework examine how publics respond to information that is intended to encourage social acceptance of such technologies (e.g. Phillips and Dickie 2015; Whitmarsh *et al.* 2015). In social psychology, for example, intense studies have scrutinised the effects of confirmation bias: the influence of prior held beliefs about something on the processing and evaluation of new information. Confirmation bias thus refers to the way in which information is remembered, sought, and interpreted such that it confirms previously held beliefs about something (Oswald and Grosjean 2004, p. 79). Research on confirmation bias in social psychology shows that prior held beliefs about something influences information seeking (Schulz-Hardt *et al.* 2000), the evaluation of information (Lord *et al.* 1979; Edwards and Smith 1996; Fischer *et al.* 2011) and the ability to remember information (Stangor and McMillan 1992), and it does so in such a way that initial beliefs are often maintained. This is not to imply a deterministic effect, but nonetheless an influence.

More generally, Leviston *et al.* (2013) point to the need to account for biased processing of information due to other psychological mechanisms, e.g. the influence of risk frames (contextual information that has implications for risk perceptions) or in-group/out-group differences such as the fundamental attribution error. The latter concerns situations in which consequences are attributed to individuals' internal characteristics (such as personality traits), without considering the specific external circumstances that were also influential (Jones and Harris 1967).[2]

Why have psychological perspectives been so little used?

In this section, we suggest that ontological and methodological challenges may be part of the reason why the psychological literature has been used rather sparingly in the sociotechnical sustainability transitions literature to date.

Experimental and variance-based research designs

Much of the behaviourally-focused environmental psychology literature that addresses issues relevant to energy transition processes has mostly analysed discrete factors such as attitudes, norms and control beliefs, and has paid less attention to broader, societal contexts. This reflects a primary interest in the structure of psychological factors, i.e. the relationship of one variable to another (for example, in terms of the direction of causality and the strength of that relationship). Such studies typically consist of experimental research designs (Aronson *et al.* 2014) that deliberately establish controlled experimental conditions (e.g. identical, careful question phrasing and information provision for each individual respondent). Moreover, the relationships between psychological factors per se and in relation to phenomena are typically examined statistically.

Hence, Grin *et al.* (2010) distinguish at a meta-level between variance and process theories as different *approaches to knowledge*. Variance theory aims to explain the variation or change in an outcome (the dependent variable) as a result of influences from causal factors (independent variables). As stated, this involves controlled conditions, assumptions of entities with fixed natures (under the same conditions) and aims at widely generalisable laws. In contrast, process theory seeks to understand patterns in events, their sequencing and the roles of actors therein (Abbott 1992; Poole *et al.* 2000; Grin *et al.* 2010). Abbott (1992) rehearses the various critiques of the assumptions underlying variable-based analysis.

As Grin *et al.* (2010) observe, theories such as the MLP and the institutional and systems perspectives favoured in sociotechnical transitions research, are processual with methodologies to match (often, small numbers of case studies, narrative and temporal, event-based detail). These authors argue that variance-theory methods have limited utility in this context because transitions processes are large scale, long term and hence relatively rare, thus precluding statistical analysis (Grin *et al.* 2010). They also argue that transitions involve complex dynamics that are difficult to explain in terms of cause-and-effect relations (Grin *et al.* 2010). It may well be that an emphasis on identifying any consistent *conditions* for change and stasis would be more fruitful. Nonetheless, variance-based methods are used across the social sciences – in politics, governance, organisational studies, entrepreneurship and so on (Sabherwal and Robey 1995; Van De Ven and Poole 2005). What matters is how they are used – what claims are reasonable according to which methods, and what are not.

Ontological differences: the individualistic focus of psychology

We discussed the issue of ontology in Chapter 1 and so elaborate only a little here, but the issue remains important. Psychological approaches by nature – and especially quite commonly used psychological theories such as the TpB – tend to deploy individualistic and cognitive paradigms of *what people are* (Keller *et al.* 2016). The same holds true for the use of other theoretical approaches that focus on the individual level, notably behavioural economics (Gazheli *et al.* 2015). Most psychological theories as applied to sustainability research have relied on a concept of *individual* agency (Jackson 2005). As Sorrell (2015, p. 79) observes: "the focus on autonomous decision-making by individuals neglects how preferences, attitudes, expectations and behaviours are embedded in and shaped by broader physical and social systems that both enable and constrain individual choice".

Such relative neglect of the social systems that embed individual actions is found in other areas of related research. Even in studies of sustainable consumption, where psychological theories have been used most in transitions contexts, Liu *et al.* (2017) conclude: "At the same time, some studies have tended to understand and predict sustainable consumption at the individual level, overlooking the social and situational factors influencing consumers' decision to act sustainably." Despite psychological and sociological frames of reference sometimes being combined sequentially in the context of sustainable consumption (e.g. Noppers *et al.* 2014), in sociotechnical transitions contexts the psychological–sociological divide seems to have kept the use of psychological approaches at a low level (Bögel and Upham 2018).

Summary

The premise underlying this book is that understanding sociotechnical transitions processes requires a multiplicity of analytic terms and perspectives. Indeed, this is already recognised: much of the contemporary sociotechnical sustainability transitions literature consists of authors bringing insights from geography, politics, governance and other disciplines.

Agency in the sense of actor behaviour in the sociotechnical sustainability transitions literature has mostly been addressed in collective contexts and at collective levels, with relative neglect of the psycho-social processes of both individuals per se and individuals in their social or societal context. In this chapter, we have documented and discussed the ways in which actor psychology is described and theorised in the literature, both implicitly and explicitly. While actor motivation and behaviour are often implicitly referred to, these are rarely theorised explicitly using psychological concepts. So far, only a small set of psychological theories has been applied to transitions research.

We have discussed possible reasons for the lack of integration of psychological theories in sociotechnical transitions thinking, and variance-based methods, individualistic ontologies and the predominance of rational choice models in contemporary behavioural psychology are suggested as possible factors. Nonetheless, we have argued that the limited range of social psychological theories that have been used in sociotechnical transitions research to date does have substantial potential. In the next chapter we look at one way of doing this while keeping close ontological coherence with theory and frameworks such as Geels' (2002) MLP, which rest explicitly or implicitly on ideas of structuration. In fact, the methodological bracketing approach that we also refer to in Chapter 4 is, in principle, ontologically flexible and can be used to justify the juxtaposition of a wide variety of types of analysis.

Notes

1 Giddens' (1984) structuration theory, which also includes an emphasis on practice, affords individuals more conscious, deliberative agency as related to "rules and resources" – although see Lizardo (2010) regarding differences between structure posited as ontologically extant (i.e. real) and structure as an analytic construct intended to explain regularities (such as the idea of a 'system').
2 This sums up the key limitations of psychological approaches to agency; the inverse applies to sociological approaches. Our premise is that use of both, sequentially or simultaneously, should provide greater insight.

Bibliography

Abbott, A. 1992. "From causes to events: Notes for narrative positivism". *Sociological Methods & Research* 20:428–55. doi:10.1177/0049124192020004002.

Adil, A. M. and Ko, Y. 2016. "Socio-technical evolution of decentralized energy systems. A critical review and implications for urban planning and policy". *Renewable and Sustainable Energy Reviews* 57:1025–37. doi:10.1016/j.rser.2015.12.079.

Ajzen, I. 1991. "The theory of planned behaviour". *Organizational Behavior and Human Decision Processes* 50(2):179–211. doi:10.1016/0749-5978(91)90020-T.

Ajzen, I. 2011. "The theory of planned behaviour. Reactions and reflections". *Psychology & Health* 26(9):1113–27. doi:10.1080/08870446.2011.613995.

Aldous, D. 2014. "Understanding the complexity of the lived experience of foundation degree sport lecturers within the context of further education". *Sport, Education and Society* 19(4):472–88. doi:10.1080/13573322.2012.674506.

Amenta, E. and Ramsey, K. M. 2010. "Institutional theory". In: E. Amenta, K. Nash and A. Scott (eds.), *Handbook of Politics*. Chichester, UK: John Wiley & Sons, pp. 15–39.

Aronson, E., Wilson, T. and Akert, R. 2014. *Sozialpsychologie* [Social Psychology]. Hallbergmoos, Germany: Pearson Studium.

Atkinson, J. W. 1957. "Motivational determinants of risk taking behaviour". *Psychological Review* 64:359–72.

Audet, R. and Guyonnaud, M-F. 2013. "Transition in practice and action in research. A French case study in piloting eco-innovations". *Innovation* 26(4):398–415. doi:10.1080/13511610.2013.850019.

Axsen, J. and Kurani, K. S. 2014. "Social influence and proenvironmental behaviour: The reflexive layers of influence framework". *Environment and Planning B: Planning and Design* 41(5):847–62. doi:10.1068/b38101.

Batel, S., Castro, P., Devine-Wright, P. and Howarth, C. 2016. Developing a critical agenda to understand pro-environmental actions: Contributions from social representations and social practices theories". *Wiley Interdisciplinary Reviews: Climate Change* 7(5):727–45. doi:10.1002/wcc.417.

Beckers, T., Siegers, P. and Kuntz, A. 2012. "Congruence and performance of value concepts in social research". *Survey Research Methods* 6(1):13–24.

Bijker, W. E. 2009. "Social construction of technology". In: J. K. Olsen, S. A. Pedersen and V. H. Hendricks (eds.), *A Companion to the Philosophy of Technology*. Chichester, UK: Wiley-Blackwell, pp. 88–94.

Bijker, W. E., Hughes, T. P. and Pinch, T. J. 1987. *The Social Construction of Technological Systems*. Cambridge, MA: MIT Press.

Blumer, H. 1969. *Symbolic Interactionism: Perspective and Method*. Englewood Cliffs, NJ: Prentice-Hall.

Bögel, P. and Upham, P. 2018. "Role of psychology in sociotechnical transitions studies: Review in relation to consumption and technology acceptance". *Environmental Innovation and Societal Transitions* 28:122–36. doi:10.1016/j. eist.2018.01.002.

Bourdieu, P. 1986. "The forms of capital". In: J. G. Richardson (ed.), *Handbook of Theory and Research for the Sociology of Education*. New York: Greenwood, pp. 241–58.

Bowerman, T. 2014. "How much is too much? A public opinion research perspective". *Sustainability: Science, Practice, and Policy* 10(1):14–28.

Bristow, G. and Healy, A. 2014. "Regional resilience. An agency perspective". *Regional Studies* 48(5):923–35. doi:10.1080/00343404.2013.854879.

Brones, F. A., Carvalho, M. M. D. and Zancul, E. D. S. 2017. "Reviews, action and learning on change management for ecodesign transition". *Journal of Cleaner Production* 142:8–22. doi:10.1016/j.jclepro.2016.09.009.

Brown, H. S. and Vergragt, P. J. 2016. "From consumerism to wellbeing: Toward a cultural transition?". *Journal of Cleaner Production* 132:308–17. doi:10.1016/ j.jclepro.2015.04.107.

Büscher, C. and Sumpf, P. 2015. "'Trust' and 'confidence' as socio-technical problems in the transformation of energy systems". *Energy, Sustainability and Society* 5(1):1–13. doi:10.1186/s13705-015-0063-7.

Chaiklin, H. 2011. "Attitudes, behavior, and social practice". *Journal of Sociology & Social Welfare* 38(1):31–54. Available at: https://scholarworks.wmich.edu/ jssw/vol38/iss1/3.

Chatterton, P. 2016. "Building transitions to post-capitalist urban commons". *Transactions of the Institute of British Geographers* 41(4):403–15. doi:10.1111/ tran.12139.

Clayton, S., Devine-Wright, P., Stern, P. C., Whitmarsh, L., Carrico, A., Steg, L., Swim, J. and Bonnes, M. 2015. "Psychological research and climate change". *Nature Climate Change* 5:640–6.

Cohen, M. J. 2013. "Collective dissonance and the transition to post-consumerism". *Futures* 52:42–51. doi:10.1016/j.futures.2013.07.001.

Dawson, N. and Martin, A. 2015. "Assessing the contribution of ecosystem services to human wellbeing: A disaggregated study in western Rwanda". *Ecological Economics* 117:62–72. doi:10.1016/j.ecolecon.2015.06.018.

Devine-Wright, P. 2009. "Rethinking NIMBYism: The role of place attachment and place identity in explaining place-protective action". *Journal of Community & Applied Social Psychology* 19(6):426–41. doi:10.1002/casp.1004.

Edwards, K. and Smith, E. E. 1996. "A disconfirmation bias in the evaluation of arguments". *Journal of Personality and Social Psychology* 71(1):5–24. doi:10.1037//0022-3514.71.1.5.

Emery, F. E. and Trist, E. L. 1960. "Socio-technical systems". In: C. W. Churchman and M. Verhulst (eds.), *Management Science Models and Techniques*. Oxford: Pergamon, pp. 83–97.

Evensen, D. and Stedman, R. 2016. "Scale matters: Variation in perceptions of shale gas development across national, state, and local levels". *Energy Research & Social Science* 20:14–21. doi:10.1016/j.erss.2016.06.010.

Feola, G. and Nunes, R. 2014. "Success and failure of grassroots innovations for addressing climate change: The case of the Transition Movement". *Global Environmental Change* 24(1):232–50. doi:10.1016/j.gloenvcha.2013.11.011.

Festinger, L. 1957. *A Theory of Cognitive Dissonance*. Stanford, CA: Stanford University Press.

Fischer, P., Lea, S., Kastenmüller, A., Greitemeyer, T., Fischer, J. and Frey, D. 2011. "The process of selective exposure: Why confirmatory information search weakens over time". *Organizational Behavior and Human Decision Processes* 114(1):37–48. doi:10.1016/j.obhdp.2010.09.001.

Folke, C. 2016. "Resilience (republished)". *Ecology and Society* 21(4):44. doi:10.5751/ES-09088-210444.

Gazheli, A., Antal, M. and van den Bergh, J. 2015. "The behavioral basis of policies fostering long-run transitions: Stakeholders, limited rationality and social context". *Futures* 69:14–30. doi:10.1016/j.futures.2015.03.008.

Geels, F. W. 2002. "Technological transitions as evolutionary reconfiguration processes: A multi-level perspective and a case-study". *Research Policy* 31:1257–74.

Giddens, A. 1984. *The Constitution of Society*. Berkeley, CA: University of California Press.

Goedkoop, F. and Devine-Wright, P. 2016. "Partnership or placation? The role of trust and justice in the shared ownership of renewable energy projects". *Energy Research & Social Science* 17:135–46. doi:10.1016/j.erss.2016.04.021.

Grin, J., Rotmans, J., Schot, J., Geels, F. and Loorbach, D. 2010. *Transitions to Sustainable Development: New Directions in the Study of Long Term Transformative Change*. New York: Routledge.

Habermas, J. 1996. *Between Facts and Norms: Contributions to a Discourse Theory of Law and Democracy*. Cambridge: Polity Press.

Hassink, J., Grin, J. and Hulsink, W. 2013. "Multifunctional agriculture meets health care: Applying the multi-level transition sciences perspective to care farming in the Netherlands". *Sociologia Ruralis* 53(2):223–45. doi:10.1111/j.1467-9523.2012.00579.x.

Hynes, M. 2016. "Developing (tele)work? A multi-level sociotechnical perspective of telework in Ireland". *Research in Transportation Economics* 57:21–31. doi:10.1016/j.retrec.2016.06.008.

Jackson, T. 2005. "Motivating sustainable consumption. A review of evidence on consumer behaviour and behavioural change". A report to the Sustainable Development Research Network, Centre for Environmental Strategy, University of Surrey, UK. Available at: http://sustainablelifestyles.ac.uk/sites/default/files/motivating_sc_final.pdf.

Jonas, E., Schulz-Hardt, S., Frey, D. and Thelen, N. 2001. "Confirmation bias in sequential information search after preliminary decisions: An expansion of dissonance theoretical research on selective exposure to information". *Journal of Personality and Social Psychology* 80(4):557–71.

Jones, E. E. and Harris, V. A. 1967. "The attribution of attitudes". *Journal of Experimental Social Psychology* 3(1):1–24. doi:10.1016/0022-1031(67)90034-0.

Keller, M., Halkier, B. and Wilska, T-A. 2016. "Policy and governance for sustainable consumption at the crossroads of theories and concepts". *Environmental Policy and Governance* 26(2):75–88. doi:10.1002/eet.1702.

Kent, J. L. 2015. "Still feeling the car – The role of comfort in sustaining private car use". *Mobilities* 10(5):726–47. doi:10.1080/17450101.2014.944400.

Koutiva, I., Gerakopoulou, P., Makropoulos, C. and Vernardakis, C. 2017. "Exploration of domestic water demand attitudes using qualitative and quantitative social research methods". *Urban Water Journal* 14(3):307–14. doi:10.1080/1573062X.2015.1135968.

Leviston, Z., Browne, A. L. and Greenhill, M. 2013. "Domain-based perceptions of risk: A case study of lay and technical community attitudes toward managed aquifer recharge". *Journal of Applied Social Psychology* 43(6):1159–76. doi:10.1111/jasp.12079.

Lilliestam, J., Ellenbeck, S., Karakosta, C. and Caldés, N. 2016. "Understanding the absence of renewable electricity imports to the European Union". *International Journal of Energy Sector Management* 10(3):291–311. doi:10.1108/IJESM-10-2014-0002.

Liu, Y., Qu, Y., Lei, Z. and Jia, H. 2017. "Understanding the evolution of sustainable consumption research". *Sustainable Development* 25(5):414–30. doi:10.1002/sd.1671.

Lizardo, O. 2010. "Beyond the antinomies of structure: Levi-Strauss, Giddens, Bourdieu, and Sewell". *Theory and Society*, 39(6):651–88. doi:10.1007/s11186-010-9125-1.

Lord, C. G., Ross, L. and Lepper, M. R. 1979. "Biased assimilation and attitude polarization: The effects of prior theories on subsequently considered evidence". *Journal of Personality and Social Psychology* 37:2089–109.

Ma, G., Andrews-Speed, P. and Zhang, J. 2013. "Chinese consumer attitudes towards energy saving: The case of household electrical appliances in Chongqing". *Energy Policy* 56:591–602. doi:10.1016/j.enpol.2013.01.024.

Macmillan, A., Davies, M., Shrubsole, C., Luxford, N., May, N., Chiu, L. F., Trutnevyte, E., Bobrova, Y. and Chalabi, Z. 2016. "Integrated decision-making about housing, energy and wellbeing: A qualitative system dynamics model". *Environmental Health* 15(Suppl 1):S37. doi:10.1186/s12940-016-0098-z.

Mangone, E. 2017. *Social and Cultural Dynamics: Revisiting the work of Pitrim A. Sorokin*. Berlin: Springer.

Martin, C. J. and Upham, P. 2016. "Grassroots social innovation and the mobilisation of values in collaborative consumption: A conceptual model". *Journal of Cleaner Production* 134:204–13. doi:10.1016/j.jclepro.2015.04.062.

Miller, T. R., Wiek, A., Sarewitz, D., Robinson, J., Olsson, L., Kriebel, D. and Loorbach, D. 2014. "The future of sustainability science: A solutions-oriented research agenda". *Sustainability Science* 9(2):239–46. doi:10.1007/s11625-013-0224-6.

Missimer, M., Robèrt, K-H. and Broman, G. 2017. "A strategic approach to social sustainability – Part 1: Exploring the social system". *Journal of Cleaner Production* 140:32–41. doi:10.1016/j.jclepro.2016.03.170.

Moscovici, S. 1988. "Notes towards a description of social representations". *European Journal of Social Psychology* 18:211–50.

Murphy, J. and Smith, A. 2013. "Understanding transition–periphery dynamics: Renewable energy in the highlands and Islands of Scotland". *Environment and Planning A* 45(3):691–709. doi:10.1068/a45190.

Nail, P. R. and Boniecki, K. A. 2011. "Inconsistency in cognition: Cognitive dissonance". In: D. Chadee (ed.), *Theories in Social Psychology*. Oxford: Wiley-Blackwell, pp. 44–71.

Nilsson, M. and Nykvist, B. 2016. "Governing the electric vehicle transition – Near term interventions to support a green energy economy". *Applied Energy* 179:1360–71. doi:10.1016/j.apenergy.2016.03.056.

Noppers, E. H., Keizer, K., Bolderdijk, J. W. and Steg, L. 2014. "The adoption of sustainable innovations: Driven by symbolic and environmental motives". *Global Environmental Change* 25(1):52–62. doi:10.1016/j.gloenvcha.2014.01.012.

Nye, M., Whitmarsh, L. and Foxon, T. 2010. "Sociopsychological perspectives on the active roles of domestic actors in transition to a lower carbon electricity economy". *Environment and Planning A* 42(3):697–714. doi:10.1068/a4245.

Oswald, M. E. and Grosjean, S. 2004. "Confirmation bias". In: R. Pohl (ed.), *Cognitive Illusions: A Handbook on Fallacies and Biases in Thinking, Judgement and Memory*. Hove, UK: Psychology Press, pp. 79–96.

Phillips, M. and Dickie, J. 2015. "Climate change, carbon dependency and narratives of transition and stasis in four English rural communities". *Geoforum* 67:93–109. doi:10.1016/j.geoforum.2015.10.011.

Pinch, T. J. and Bijker, W. E. 1984. "The social construction of facts and artefacts: Or how the sociology of science and the sociology of technology might benefit each other". *Social Studies of Science* 14(3):399–441.

Poole, M. Scott, Van de Ven, Andrew H., Dooley, Kevin and Holmes, Michael E. 2000. *Organizational Change and Innovation Processes: Theory and Methods for Research*. New York: Oxford University Press.

Puhe, M. and Schippl, J. 2014. "User perceptions and attitudes on sustainable urban transport among young adults: Findings from Copenhagen, Budapest and Karlsruhe". *Journal of Environmental Policy and Planning* 16(3):337–57. doi:10.1080/1523908X.2014.886503.

Rip, A., Misa, T. J. and Schot, J. (eds.) 1995. *Managing Technology in Society: The Approach of Constructive Technology Assessment*. London and New York: Pinter.

Rivers, W. H. R. 1916. "Sociology and psychology". *The Sociological Review* 9:1–13.

Ruppert-Winkel, C., Hussain, W. and Hauber, J. 2016. "Understanding the regional process of energy transition in Marin County, California: Applying a three-phase-model based on case studies from Germany". *Energy Research & Social Science* 14:33–45. doi:10.1016/j.erss.2016.01.003.

Sabherwal, R. and Robey, D. 1995. "Reconciling variance and process strategies for studying information system development". *Information Systems Research* 6:303–27. doi:10.1287/isre.6.4.303.

Sarrica, M., Brondi, S., Cottone, P. and Mazzara, B. M. 2016. "One, no one, one hundred thousand energy transitions in Europe: The quest for a cultural approach". *Energy Research & Social Science* 13:1–14. doi:10.1016/j.erss.2015.12.019.

Schulz-Hardt, S., Frey, D., Lüthgens, C. and Moscovici, S. 2000. "Biased information search in group decision making". *Journal of Personality and Social Psychology* 78(4):655–69. doi:10.1037//0022-3514.78.4.655.

Scott, T. J., Politte, A., Saathoff, S., Collard, S., Berglund, E., Barbour, J. and Sprintson, A. 2014. "An evaluation of the stormwater footprint calculator and the hydrological footprint residence for communicating about sustainability in stormwater management". *Sustainability: Science, Practice, and Policy* 10(2): 51–64.

Shove, E. 2010. "Beyond the ABC: Climate change policy and theories of social change". *Environment and Planning A* 42(6):1273–85. doi:10.1068/a42282.

Smith, A., Stirling, A. and Berkhout, F. 2005. "The governance of sustainable sociotechnical transitions". *Research Policy* 34:1491–510.

Sopha, B. M., Klöckner, C. A. and Hertwich, E. G. 2013. "Adoption and diffusion of heating systems in Norway: Coupling agent-based modeling with empirical research". *Environmental Innovation and Societal Transitions* 8:42–61. doi:10. 1016/j.eist.2013.06.001.

Sorrell, S. 2015. "Reducing energy demand: A review of issues, challenges and approaches". *Renewable and Sustainable Energy Reviews* 47:74–82. doi:10.1016/ j.rser.2015.03.002.

Stangor, C. and McMillan, D. 1992. "Memory of expectancy-congruent and expectancy-incongruent information: A review of the social and social developmental literatures". *Psychological Bulletin* 111:42–61.

Stedman, R. C. 2016. "Subjectivity and social-ecological systems: A rigidity trap (and sense of place as a way out)". *Sustainability Science* 11(6):891–901. doi:10.1007/s11625-016-0388-y.

Stephenson, J., Barton, B., Carrington, G., Gnoth, D., Lawson, R. and Thorsnes, P. 2010. "Energy cultures: A framework for understanding energy behaviours". *Energy Policy* 38:6120–9. doi:10.1016/j.enpol.2010.05.069.

Stephenson, J., Hopkins, D. and Doering, A. 2015. "Conceptualizing transport transitions: Energy Cultures as an organizing framework. *Wiley Interdisciplinary Reviews: Energy and Environment* 4(4):354–64. doi:10.1002/wene.149.

Süsser, D., Döring, M. and Ratter, B. M. W. 2017. "Harvesting energy: Place and local entrepreneurship in community-based renewable energy transition". *Energy Policy* 101:332–41. doi:10.1016/j.enpol.2016.10.018.

Thoits, P. A. 1995. "Social psychology: The interplay between sociology and psychology". *Social Forces* 73(4):1231–43.

Upham, P., Lis, A., Riesch, H. and Stankiewicz, P. 2015. "Addressing social representations in socio-technical transitions with the case of shale gas". *Environmental Innovation and Societal Transitions* 16:120–41. doi:10.1016/j.eist.2015.01.004.

Van De Ven, A. H. and Poole, M. S. 2005. "Alternative approaches for studying organizational change". *Organization Studies* 26:1377–404. doi:10.1177/017084 0605056907.

Webb, Thomas L., Sniehotta, Falko F. and Michie, Susan 2010. "Using theories of behaviour change to inform interventions for addictive behaviours". *Addiction* 105(11):1879–92. doi:10.1111/j.1360-0443.2010.03028.x.

Wesseling, J. H. 2016. "Explaining variance in national electric vehicle policies". *Environmental Innovation and Societal Transitions* 21:28–38. doi:10.1016/j. eist.2016.03.001.

Whitmarsh, L. 2012. "How useful is the multi-level perspective for transport and sustainability research?" *Journal of Transport Geography* 24:483–7. doi: 10.1016/j.jtrangeo.2012.01.022.

Whitmarsh, L. and Lorenzoni, I. 2010. "Perceptions, behavior and communication of climate change". *Wiley Interdisciplinary Reviews: Climate Change* 1:158–61. doi:10.1002/wcc.7.

Whitmarsh, L., Nash, N., Upham, P., Lloyd, A., Verdon, J. P. and Kendall, J-M. 2015. "UK public perceptions of shale gas hydraulic fracturing: The role of audience, message and contextual factors on risk perceptions and policy support". *Applied Energy* 160:419–30. doi:10.1016/j.apenergy.2015.09.004.

Yun, S. and Lee, J. 2015. "Advancing societal readiness toward renewable energy system adoption with a socio-technical perspective". *Technological Forecasting and Social Change* 95:170–81. doi:10.1016/j.techfore.2015.01.016.

Zehr, S. 2015. "The sociology of global climate change". *Wiley Interdisciplinary Reviews: Climate Change* 6(2):129–50. doi:10.1002/wcc.328.

4 Strong structuration as an integrating framework for psychological and sociological perspectives

Introduction

In this chapter, we develop a theoretical framework that allows for the integration of psychological and sociological perspectives on individual 'responses' to lower carbon energy technology. We do so drawing on Upham *et al.* (2017), who apply Stones' (2005) *strong structuration* framework. The framework that we propose can accommodate profound differences in research perspectives and approaches. It does this through the use of *methodological bracketing* combined with a structuration based rationale for juxtaposing a sequence of studies that uses (if appropriate) different types of research approach, each of which sheds a different type of light on the phenomena in question. Here, those phenomena concern the many different types of processes involved in sociotechnical change and stasis, including in relation to energy systems. While many different disciplines have much to say about those dynamics, our focus here is on the contribution of social psychology; we have chosen a structuration based framework to be consistent with the structuration premise of Geels' (2002) multi-level perspective (MLP), given the frequent reference to this heuristic in the sociotechnical transitions literature.

Box 4.1 Methodological bracketing

The term 'methodological bracketing' (MB) is used in qualitative research in a variety of ways. In this chapter, we refer to Giddens' use of the concept as a means of operationalising structuration theory in empirical research. In a summary of the agency–structure debate in relation to Giddens, Feeney and Pierce (2016) suggest Giddens separated (i.e. bracketed out): (i) agents' internal skills, awareness and knowledgeability from (ii) the rules and resources of the institution (meaning the social context) in which those agents act. In other words, relative to agents, Giddens analytically created internal and external spheres that separated agency and structure. For our purpose, that which Giddens recognised as the internal sphere of the agent can be analysed psychologically, but this analytic separation means that the *connections* between the internal and external spheres of agents are not theorised (or at least are arguably under-theorised).

The value of this analytic separation depends on one's purpose, and here we are interested in the connections between the posited internal and external spheres, because we want to connect psychological aspects of agents to the sociotechnical structures that they inhabit, seek to change or seek to keep the same. Stones (2005) retains the idea of methodological bracketing, but he emphasises the value of studying "the connecting tissue between the two brackets" (Parker 2006; see also Feeney and Pierce 2016). One might argue that this takes Giddens' notion of structuration closer to Bhaskar's critical realism, in that both Stones' (2005) and Bhaskar's (e.g. 2014) ontologies acknowledge the co-existence of different types of processes in a given situation. Acknowledgement of this strengthens the case for interdisciplinary or multi-disciplinary studies, if one's objective is to gain a fuller understanding of a social or sociotechnical situation.

There is another, closely related and more specific, way in which the term 'methodological bracketing', which itself has many varieties, is used in the social sciences. This is bracketing for the purpose of demonstrating data validity in qualitative research, especially in phenomenology, ethnography and grounded research. MB in this sense comprises critical reflection vis-à-vis researcher positioning in the field; prior knowledge of the research topic; researcher biases and preconceptions; and the potential impact of researcher values/value-judgements, personal history and culture on the research (Gearing 2004; Tufford and Newman 2012; Chan *et al.* 2013). Here, the process of MB involves continuously 'thinking through' the possible impact of such researcher positioning during the iterative process of research. In a more abstract sense, one might say that MB includes the very conscious and deliberate acknowledgement and recognition of the researcher's epistemological and ontological stance both prior to and during the research process (see Gearing 2004). Although this understanding of the term is relevant here, it is not the primary sense in which we mean it and we refer to this additional meaning primarily for clarity.

In the first part of this chapter we set out the framework in some detail; in the second part we begin to illustrate parts of it. To illustrate and apply it fully, however, would take several chapters or even a separate book. For an example of a full application of Stones' (2005) framework, see Fjellstedt (2015), who used Stones' approach to explore the interconnected psychological and structural issues involved in a technology focused organisational change context (specifically, regarding the introduction of a new data management system in a hospital in the United States).

The strong structuration framework and conjunctural knowledge

Stones (2005) strong structuration approach is an extension of Giddens' (1984) approach to structuration. It includes actor psychology in a four-fold

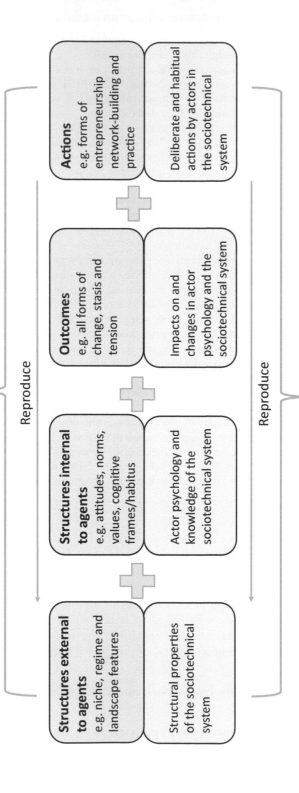

Figure 4.1 Application of the strong structuration framework (Stones 2005) to elements of sociotechnical change.

account of how actors change and maintain different types of social (including organisational and institutional) structures. The integrated framework that we propose connects the psychological to the basic structural elements of Geels' (2002) MLP (see Figure 4.1). By means of this integrated framework, we show how actor perceptions of an exemplar technology (home fuel cells for heat and power) are conditioned by the actors' own experiences and situations, and how these perceptions have implications for both their own actions and for those of others (Upham *et al.* 2017).

We also highlight the extent to which actor views and technology-related perceptions can focus on the anticipated actions of other actors within the systems, and on the way in which this mutual assessment and observation forms a significant part of what Stones (2005) characterises as "conjunctural" knowledge. In this context, conjunctural knowledge is knowledge of specific causes and trends operating in a given moment, as perceived by the individual concerned. We suggest that strong structuration (Stones 2005) is well-suited to the role of connecting psychological and sociotechnical accounts through its broad scope, which extends from individual experience to system- or structure-level processes.

From a structuration perspective, social structures exist and function through agents' actions, and the agents assign specific roles and meanings to those structures (Grin *et al.* 2010, p. 233). Giddens' (1984) theory of *agency–structure as practice* constituted a seminal insight into the relationship between agency and structure via three main elements: signification, relating to discourse; legitimation, relating to values, norms and standards; and domination, relating to control over resources. Yet, Giddens' structuration theory has also been critiqued for its relatively abstract nature, and for neglecting the situated detail of actors, particularly actor psychology (such as motivations, attitudes, beliefs and knowledge) and their social, cultural and political contexts (including organisational positions and roles) (Stones 2005).

Box 4.2 Giddens' structuration theory and Stones' strong structuration framework

Giddens' ambition was to create a theory that represented society and social practices, agency and structure as mutually and dynamically co-evolving; as produced and reproduced and ultimately shaped by the ongoing doings and actions of active subjects, i.e. by human agency across space and time (Giddens 1984). Accordingly, Giddens' structuration theory "is predicated on two assumptions: (1) social structure and agency are in practice not separable (except in the process of theoretical analysis) and (2) the relationships between agency and social structures are mutually reciprocating" (Schwandt and Szabla 2013, p. 3). However, Giddens' structuration theory has nonetheless been critiqued for paying too little attention to agency and structure, epistemology and research methodology (Schwandt and Szabla 2013).

Stones' strong structuration framework builds upon the work of Giddens, allowing for the "reciprocal interplay between the micro- and macro-levels while focusing on the centrality of human agency" (Bodolica *et al.* 2016, p. 796). Stones' theory draws upon the concept of 'ontology-in-situ' (i.e. entities and actions exist in their original place of occurrence) and considers human agents as linked in dynamic networks of position practices, here referring to the interdependent practices and relations of agents connected within a given context or field. The strong structuration model includes:

> "external structures" as conditions of action (i.e. context); "internal structures" which are present in the mind of the agent (i.e. knowledge and schemata with which to interpret meanings); "active agency" that includes specific practices enacted by agents in a given context; and "outcomes" as the result of human agents' enactments, generating variance in external and internal structures.
>
> (Bodolica *et al.* 2016, p. 7)

Stones (2005, p. 88) defines general dispositional knowledge as taken-for-granted skills and dispositions,

> encompassing generalized worldviews, cultural schemata, classifications, typified recipes of action, deep binary frameworks of signification, habits of speech and gestures, and methodologies for adapting this generalized knowledge to a range of particular practices in particular locations in time and space.

By adding (or at least highlighting) structures internal to agents to structuration theory, Stones' strong structuration framework adds both specificity and complexity to how we understand the dynamics of human interactions. Specifically, it allows for the analysis of human agency at a psychological as well as at a social level; for the analysis of dynamic relations of knowledge, status, power, judgemental processes and cognitive processes within a given societal context (Schwandt and Szabla 2013; Bodolica *et al.* 2016).

The 'strong structuration' framework was originally developed to explicitly account for agents' psychology, in recognition of the ways in which this can influence and be influenced by organisational and wider structural change (Stones 2005). In fact, Geels and Schot (2010) do briefly acknowledge Stones' (2005) structuration approach as relevant for the analysis of the motivations, perceptions, aims and interests of actors in local projects and niches as addressed in the MLP (Geels and Schot 2010). Extending Giddens' theory, Stones' 'strong structuration' posits four elements that are intended to represent key aspects of structuration more fully than Giddens' (1984) original account: (1) external structures that condition the actions of agents; (2) internal, psychological structures within agents; an assumption of (3) active, intentional agency; and (4) outcomes that are internal and/or external to the agent (Stones 2005; Fjellstedt 2015). More specifically, Stones (2005) explains the four elements as follows:

1   *External structures* are understood as separate from the agent, and they set boundary conditions; in a sociotechnical context, they include laws, regulations, formal and informal institutions, policies and organisations.

2   *Internal structures* are divided into two classes: those that are *general-dispositional*, including habitus,[1] norms, values,[2] attitudes, worldview, etc., depending on one's analytic frame, and those that are *conjunctural*, relating to agents' knowledge and understanding of their immediate and wider social and cultural context. While these two classes are analytically distinct, in practice they may overlap.

3   *Actions* (active agency) relate to processes of thoughtful and sometimes strategic action: in this context, active agency would include all forms of entrepreneurship, lobbying, coalition-building, resistance and 'mindful deviation' (conscious innovation) (Garud and Karnøe 2001) in general.

4   *Outcomes* are any consequences of the foregoing for the agents and their structural context, here including all forms of sociotechnical pathway and the tensions that these comprise.

Application of the strong structuration framework to sociotechnical change

Here, we use the three-level conceptualisation of the external structures of interest within sociotechnical systems as set out in Geels' (2002) MLP. Stones' (2005) strong structuration approach is then placed in relation to these three levels and thus also to sociotechnical sustainability transitions features and processes more generally (see Figure 4.1). Overall, the integrative framework supports the use of different types of analysis, i.e. different methodologies and disciplinary perspectives. As Figure 4.1 implies, the research design invited by this framework is a sequence of closely related studies that reveal the connections between individual and structure level processes in much more detail than can be achieved with a mono-disciplinary research approach.

Structuration or critical realism?

At its core, Stones' (2005) strong structuration framework consists of a conceptual frame that supports the formal juxtaposition and connection of differing types and levels of analysis. This has implications for the analysis of social – and here sociotechnical – phenomena, mostly because of the way in which the framework supports extended, interconnected and layered analysis, including the use of analytic approaches that have differing ontologies. Studies of behaviour or agency retaining framework-level ontological consistency with the MLP (Geels 2002) would use a structuration perspective to capture actor/agency perspectives. As stated in Chapter 1, in the MLP the three levels of interacting phenomena, the niche, regime and landscape, represent three different degrees of structuration (Geels and Schot 2007; Grin *et al.* 2010, pp. 44–7). Moreover, Giddens' assumption of active, reflexive agents is consistent with the proposition that agents can act and thus choose between alternative structures (Coad *et al.* 2015), a feature that Stones' (2005) framework also emphasises.

However, as observed in Chapter 1, there is a legitimate and we think valuable debate around the extent to which structuration acts as a constraint on the analysis of sociotechnical transitions and the relative value of a critical realist ontology. Hence Svensson and Nikoleris (2018), drawing on Archer (1995), emphasise that Giddens' 'rules and resources' based description of structure says little about the differences between people in access to knowledge and resources; these differences may arise from social positions, access to material resources or other factors. Svensson and Nikoleris (2018) also argue that the socio-cognitive rules[3] (of Giddens), those that in the MLP instantiate a regime, may change without the fundamental structure of a regime changing, or that the same rule changes lead to different changes in different contexts.

Sorrell (2018) argues that, in practice, many authors (including Geels) have been flexible in their implicit ontological assumptions when analysing sociotechnical systems. Nonetheless, the above issues remain as limitations of the structuration base of the MLP (Sorrell 2018). While Sorrell's preferred option is critical realism, Stones (2005) strong structuration framework incorporates more of the issues (including power, politics and subjective experience) that the MLP arguably downplays. It does not, though, 'fix' the fundamental feature of structuration that critical realists object to, namely, that viewing social structure as an outcome of socio-cognitive rules neglects the way in which structure can pre-exist and shape those rules. As an explanation of social structure, Giddens' (1984) structuration theory arguably gives too much power to the socio-cognitive, too little power to the external reality of the social system and reduces social structure to the outcome of social practices (Archer 1995). Stones (2005) structuration framework does facilitate a focus on actors and agency: topics that the sociotechnical literature has been critiqued for neglecting. It does create opportunities for the combination of research insights from multiple fields and encourages sociotechnical transitions researchers to reflect on the nature of social structure and to use the term 'sociotechnical' less blithely. It cannot, though, resolve the limitations of Giddens' approach to structuration per se.

In the next sections we illustrate how the integrated framework, also set out in Upham *et al.* (2017), can help connect – in this case – personal experience to sociotechnical structures, and we do this via the example of stakeholder opinion of hydrogen fuel cell (HFC) applications. Informants for this qualitative case study are from the UK, Germany and Spain. The case study draws on Upham *et al.* (2017). The HFC applications are for use at different scales and in different contexts, for example in individual households or for the heat and power supply of larger communal buildings.

Case study: hydrogen applications for heat and power

National policy and technology context: the UK, Germany and Spain

Residential-scale, fuel cell based, combined heat and power (CHP) units are among the micro-level options for generating lower carbon heat and electricity, particularly when the energy vector (e.g. hydrogen) is produced using renewable energy. Home fuel cells should have no harmful emissions, and they have reached a commercialisation stage in several countries (e.g. Japan, South Korea and the United States) (Ammermann *et al.* 2015). Although the technology is expected to remain comparatively expensive in the short and medium term, in the long term, home fuel cells are perceived as having mass-market potential – though only if costs can be reduced (ibid.). While scaling up production could potentially reduce production costs, this also requires an increased demand for the technologies, which in turn requires public policy support in terms of, for example, subsidies to reduce the cost of the initial investments (ibid.).

Currently, in Germany, Spain and the UK electricity for households is supplied almost universally by national networks, although heat is supplied through a variety of means, with more diversity in Germany than in the UK and Spain. In the UK, domestic heating is still largely based on natural gas and, as of 2015, approximately 90 per cent of UK households had their own boilers (DCLG 2017). Until relatively recently, renewable energy provision in the UK was encouraged (including for heating), but currently the policy direction in this regard is less clear. In 2013, the UK issued the equivalent of a national heating strategy, "The future of heating: Meeting the challenge" (DECC 2013). This strategy makes 42 direct references to hydrogen, and it views domestic and industrial demand and supply in relation to one another, i.e. in systems terms. In the strategy, one of the ambitions is to make use of waste heat: this is heat that is generated, primarily by industrial processes, that might be made use of but which is currently disposed of as, for example, steam to the atmosphere. A highly conservative estimate of this wastage for the UK equates to the lost opportunity to save around one million tonnes of CO_2 emissions per year. This equates to the annual CO_2 emissions of (very roughly) 200,000 UK homes, or a modest sized town.[4]

Only a few years later, the 2016 consultative document "Heat in buildings" (DBEIS 2016) had a much narrower scope. This document focuses largely on domestic boiler design, refers to district heating only once, and not at all to hydrogen. The UK has been viewed by the European Union as falling behind on renewables targets, and in 2017 the UK was less than halfway towards meeting its target of supplying 12 per cent of heating from renewable energy sources (House of Commons Energy and Climate Change Committee 2016). Nonetheless, the carbon intensity of electricity production has fallen substantially: from over 500 g/kWh in 1990 to well below 300 g/kWh in 2016, and in the summer months this approaches 200 g/kWh (CCC 2017; Fankhauser *et al.* 2018).

By contrast, Germany has, and is implementing, a 2014 National Action Plan on Energy Efficiency (NAPE) that focuses not only on the energy efficient upgrade of buildings but also on the use of buildings and urban spaces. As a result, Germany has greatly expanded its district heating share of household heating, despite oil and gas-fired boilers being the typical heating source in single, two-family and multi-family dwellings (Federal Ministry for Economic Affairs and Energy 2015a). The result is that since 2014, 21.5 per cent of newly built homes in Germany have relied on district heating, bringing this to a total of 13.5 per cent of all homes. However, natural gas boilers were still installed in 49.8 per cent of newly built homes in Germany (Federal Ministry for Economic Affairs and Energy 2015b).

The policy environment for renewables in both the UK and Germany is significantly better than it is in Spain. Alonso *et al.* (2016) colourfully describe this policy change as Spain "losing the roadmap" in terms of transitions to low carbon energy supply. By 2012, Spain was a major European player vis-à-vis the implementation of wind power, solar thermal electricity and photovoltaics but, since then, all subsidies for renewable energy have disappeared and this has had a severe impact on the renewables sector in the country (ibid.). Interestingly, Alonso *et al.* (2016, p. 681) emphasise that "the marked differences in development of [renewable energy] across the Spanish regions strongly reflects the different territorial, economic and administrative circumstances in each region".

Aim of the study

The aim of the study was to provide an insight into a range of stakeholder perceptions and opinions of hydrogen and HFCs for heat and power in a European context. The chosen stakeholders/informants in this qualitative study were all professionals working with hydrogen and HFCs in different ways, and data from the qualitative interviews does suggest that perceptions and opinions of these technologies among the interviewees may be partially shaped by the policy and sociotechnical environments in which they are integrated. In this way, the data underscore the value of the integrated framework that we propose in this chapter for understanding the wider connections between structure, agency and – importantly – the dynamics of *change* in sociotechnical systems.

Method

To gather data, 145 semi-structured interviews were undertaken in Spring 2016, to document relevant aspects of informants' conjunctural knowledge and lived experience vis-à-vis the hydrogen and HFC technologies (Upham *et al.* 2017). Here we only make claims regarding the *illustration* of our theoretical points, given that we do not know (or have any definitive way of knowing) the total number of people involved in the extended innovation system of hydrogen and HFCs in Europe. Nonetheless the number of interviews undertaken was substantial for a qualitative study and the points we make below are typical of the wider group of those interviewed.

The interviews were conducted in five European Union countries to cover different levels of innovation intensity in this particular field: France, Germany, Spain, Slovenia and the UK (Upham *et al.* 2017). However, for brevity, we focus in this chapter on the UK, Germany and Spain. Most of the interviewees worked in private companies or in publicly funded research organisations; others were from government organisations and some were from other non-profit organisations. Interviewees were identified and selected from project[5] partner databases, and the overall research strategy was to obtain a spread of opinion across the hydrogen and HFC innovation community and its stakeholders. By country, the percentage of all interviewees was: Spain, 27 per cent; Germany, 23 per cent; UK, 16 per cent; France, 26 per cent; and Slovenia, 8 per cent. Although there were HFC-related innovation activities in all of the five European countries in which interviews were conducted, often funded or co-funded by the European Union, only Germany had a dedicated national hydrogen implementation plan.[6] Table 4.1 shows the organisational affiliation of the interviewees by nationality.

As we shall see, the selected quotations all highlight the connections between interviewees' opinions of hydrogen and fuel cell technologies; their organisational affiliations and roles; their national policy and economic contexts; their reflexive awareness of other actors and the possible, future actions of those actors; and the implications of all of this knowledge for the progress of HFCs. In Giddens' terms, such knowledge is a key *resource*, with knowledge of present and future *rules* (individual, organisational, governmental and beyond) also understood to guide behaviour.[7] Although it does admit to nuance, the terms of Giddens'

Table 4.1 Interviewee organisational affiliation by country (*n*)

Affiliation	Spain	Germany	UK
Local government	7	0	4
Public company	2	4	1
University or state research organisation	7	15	1
Multisector partnership	1	0	7
Government ministry or agency	8	2	4
Commercial	8	16	4
Other non-profit organisation	1	2	3
Total	34	39	24

approach to structuration are nonetheless rather few, reflecting his aim of articulating a universally applicable theoretical approach to the relationship between human agency and social structure. Stones (2005) response (strong structuration) seeks to allow a more detailed and multidisciplinary account of the relationship between agency and social structure, as mediated by the individual's *lived experience* (see e.g. Aldous 2014 for an application; Klapper and Upham 2015 for an account of lived experience). For Stones (2005) this lived experience consists of more types of phenomena than Giddens' account gives emphasis to. For our purposes here, the most important of these omissions is the internal structure of the individual, comprised of their *general-dispositional* nature (their habitus, norms, values, attitudes, worldview, etc.) and their *conjunctural knowledge* (of their immediate and wider social and cultural context). Our argument is that these internal, psychological processes and states inform individuals' action and inaction, in relationship to their social and sociotechnical structures. All of these can be summed up as their lived experience.

Stakeholder opinion

In the following sections, we present selected empirical interview data, focusing systematically on various themes of interest, divided into broad and general categories. Given the large number of interviews, we also gathered some quantitative data on the attitudes and beliefs expressed and we present this in Chapter 5. In this chapter, we focus on the qualitative experience of a number of illustrative interviewees.

Perceived strengths and weaknesses of hydrogen and HFCs

UK government ministry official

This stakeholder/informant emphasised the value of (renewable) hydrogen for incurring limited disruption of existing systems of heat provision: that is, for facilitating a low carbon energy system in a way that involves minimal inconvenience for the consumer. The interviewee's job has included commissioning studies on the decarbonisation of both the UK electricity and gas grids. He explained that over the years he has been working in this field, he has noticed that his colleagues have taken the role of hydrogen in the decarbonisation of the gas grid increasingly seriously. Their attitudes have changed from complete scepticism to the view that limited injection of renewable hydrogen to the national gas grid is an option that should be seriously examined. The informant viewed micro combined heat and power units as an option (these comprise a small reformer to separate hydrogen from natural gas fed to fuel cells), but he also considered this option to be less economically attractive for domestic users.

Overall, his experience with UK politics inclines him towards technology options that are as substitutional (drop-in) as possible – again for reasons of incurring minimal change and cost for individual consumers. More specifically, he emphasises the way in which hydrogen can be added to the existing gas grid relatively easily, acting as an alternative to piped district heat networks. This requires modest modifications to existing equipment within homes (i.e. new or modified boilers for water heating and gas hobs for cooking):

> ... with hydrogen you are talking about someone coming in one day and changing the boiler for a new boiler in much the same space, that operates in much the same way as the previous one did.... In terms of disruption, that is hugely more attractive and a much easier sell than trying to convince people to take heat pumps.... now if we are talking about heat networks, the disruption there comes in terms of years of roads being dug up as heat networks are installed, and there is quite a concern about how practical that is going to be in London.
>
> (UK, government ministry official)

The interviewee's embeddedness in and experience with this particular political system (i.e. the UK political system) informs his views on realistic policy options for increasing the hydrogen supply in the gas grid. He sees two main options: (1) a "command and control economy from a government turnaround and saying this is how it shall be", but "that is not typical of the UK"; and (2) "a roll out plan that links [all stakeholders together including] the distribution network operators over time to make that happen..." (UK, government ministry official). However, to this second option, he adds: "It is very difficult to see how you make that happen without a strong steer from government" (UK, government ministry official).

In other words, the degree of government intervention in the market that is required to make hydrogen work institutionally seems unlikely to this informant. Overall, the perspective of the informant on actual and potential sociotechnical system change explicitly recognises the significant role of end-user experiences with – and attitudes towards – the technologies in question. His perspective also emphasises the importance of the government's action – or inaction – for potential systems change. The general implication for our purposes, which we keep reiterating, is that he has come to these views through his experience and that those views have implications for the decisions and recommendations that he makes regarding policy. The sociotechnical system is made up of a large number of more or less powerful actors such as himself and their views – their individual experience, knowledge, attitudes, norms and values – all matter for system change and stasis.

Spanish academic working with hydrogen and fuel cell research

A Spanish interviewee provides insights from his long experience with hydrogen and fuel cell research. He has particular attitudes and beliefs regarding hydrogen and HFCs that are based on his experience and these have implications for his decision-making, advice and ultimately sociotechnical change. In the next chapters we show how psychological theory can be applied to such individual cases to help further explain and theorise the connections between the individual and the sociotechnical, but here we simply illustrate how individual-level processes matter for sociotechnical transitions.

Most recently, the interviewee has been working on the production and storage of hydrogen via electrolysis using surplus electricity from a wind farm. The project he is working on is sponsored by the private company who own the turbines, and this company (as others) is subject to regulated limits on their power input to the electricity grid at times of peak supply and low demand. He says that the companies he is working with are experimenting with power to gas for strategic, learning reasons, but his academic role also leans him towards a qualified (conditional) view of the current state of power to gas technology, for balancing the power in the electricity grid.[8]

> It [power to hydrogen gas] is not a mature technology, from a technical or commercial point of view. Regarding the connection with renewable energy sources, there are fundamental deficiencies caused by the dynamics of the system, in terms of electrolysers, and with respect to the durability of equipment working under a variable load ... With regard to the commercial aspects, the scaling factors are not conducive to think that it is an economically competitive technology against other competitive technologies.
>
> (Spain, researcher and academic in private company)

He continues:

> Another problem is that there is no legislation [many interviewees take the view that legislative frameworks for regulating hydrogen supply and use on a larger scale are inadequate]. We had an experience in 2008 with a wind power company. We wanted to install a demonstration facility and we collided with the legislative barrier. You go to the corresponding council and the council looks away. The city council, the regional department or the ministry do not even know where to fit a project of these characteristics, because there is no legislative basis about these technologies ... it happens in all the applications of hydrogen. Try to convince the municipal administration or the municipal technical services to install a hydrogen refuelling station in the centre of a city. Very difficult.
>
> (Spain, researcher and academic in private company)

Despite these challenges, the interviewee still views hydrogen generation and storage as preferable to other technical options in situations where the storage is large scale. One alternative is compressed air storage in underground caverns, but these are only available in particular locations. Overall, based on his experience, this informant favours renewable power to gas in principle, but he also acknowledges the technical, legislative and economic problems and challenges associated with this technology, in the short term at least.

The experience of this informant spans more than one sector: the interviewee is an academic with a history of working with commercial partners, and so his beliefs about the current state of power to gas technology are informed by his personal experience of specific technological and operational challenges. He has ample experience of regulatory uncertainty in the field. His role as an academic working for the company also involves reliably assessing market readiness for this technology. Overall, he is positive about the potential for power to compressed hydrogen gas as a method of renewable energy storage, but he also acknowledges the challenges involved.

Spanish private sector actors

Drawing upon personal observations from travelling abroad, a Spanish interviewee from a private sector company involved with HFCs reflects on the implications of different cultural norms for personal living space and power consumption:

> I was in Yokohama at the hydrogen congress in 2005. At that moment, Honda, Mitsubishi and others were rehearsing fuel cells … electric cells of 1.5 kilowatts [for home use]. In Japan, houses are almost like the living room of a Spanish house … [he is used to larger houses in Spain relative to Japan] … and imagine the situation in the US, with a cell of 1.5 kilowatts you will just illuminate the porch. For this reason it [designing home-scale HFCs] is complicated, and now the tendency goes to a self-generation photovoltaic.
>
> (Spain, private sector employee)

Another private sector informant from Spain echoes the regulatory problems that s/he has experienced with hydrogen and HFCs:

> …and then there is the problem of handling the hydrogen. The current legislation is far from clear as to hydrogen, or at least it is not easy. This is also very important. If for installing a domestic fuel cell in your house you need to study 200 laws, you are going to get crazy.
>
> (Spain, private sector employee)

Perceived expectations of systems change among interviewees

The tone was mixed among the various interviewees vis-à-vis expectations of system change to hydrogen or fuel cells: positive expectations of this change, understood as the successful implementation of the technologies, were voiced approximately as frequently as negative expectations. However, in Spain these expectations of system change were generally negative, and this was also the case when thinking longer term. Among the UK interviewees, solving the challenge of hydrogen storage was referred to as key to the take-up of stationary HFC applications for heat and power.[9] German interviewees referred to uninterruptible supply systems (back-up power systems, such as in hospitals, telecoms centres and any critical infrastructure) as a likely niche; indeed several (German and other) interviewees viewed some form of market deployment of HFCs for stationary (heat and power) applications as inevitable.

German university and government agency actors

German interviewees, who were supported by a national-level hydrogen innovation programme, had their eye on the long term for the implementation of hydrogen and HFC technologies.

> Currently, it is a "dry period" as it will take some time for the [hydrogen and HFC] market to develop. This is different from the situation in other countries, for example in Japan, where much more of these appliances are already installed and running. And the Japanese manufacturers are now also entering the German market so that this may have a negative impact on German manufacturers as they are more developed.
>
> (Germany, university research sector employee)

> …of course, it's a shame, as a young technology it is still expensive and has no market advantages. Things will arise. Overall, regarding the assessment of the technology – stationary applications – I believe that it is a very important new component, which will capture relevant market shares sooner or later.
>
> (Germany, government affiliation)

> We need a great leap forward. I question that we will manage this alone; a coordinated action is needed, by companies, investors, funding institutions and also customers, they should cooperate. Then we could succeed.
>
> (Germany, private sector affiliation)

What is perhaps most notable among the German interviews is that HFCs for stationary/heat and power use at different scales are largely taken for granted as technically plausible: here, it is mostly the *market* that is viewed as undeveloped – but these informants also expect the market to develop. The German interviewees hold several specific expectations for the technologies not mentioned by the UK or Spanish interviewees. The German informants imply that both hydrogen and HFCs could be suitable for heavy vehicles, forklifts and auxiliary power units, while HFCs could be suitable for uninterruptible power (critical systems); and that price reductions will come from economies of scale (e.g. due to experience leading to more cost effective production methods) for fuel cells. Among these interviewees, there is also a general belief that government support for these specific technologies will increase. Overall, the attitudes towards and beliefs about hydrogen and HFC technologies expressed among the German informants were cautiously positive, particularly for the longer term.

Hopefully, it should be clear through just the small number of illustrations above that actor perceptions, decisions and actions have a history based in individual experience. In the next section we discuss this further.

Discussion

Actors and system change

While the changes required for fundamental energy and wider systems change will take place at several levels (institutional, technological and others), being aware of the personal experiences and psychology of the actors in these processes will help us to understand aspects of system development, be these at posited niche, regime and landscape levels (Geels 2002), or in terms of some other conceptualisation of the sociotechnical context. Acknowledging the importance of individual level processes in and for sociotechnical systems, and particularly for systems change, enables the inclusion of corresponding accounts (psychological and sociological) of the mutual relationships between individuals and collective forms of organisation and, more generally, of how processes at both levels may influence sociotechnical structure. These accounts can relate to phenomena that include issues such as identity, learning, motivation, decision-making, affective and communicative influences, interpersonal factors, cognitive biases – and multiple other psychological processes and phenomena.

Before discussing the specific implications of the data presented above, we first make some more general points about the ontology of systems analysis. To date, the sociotechnical transitions literature has tended towards realist or positivist assumptions of system behaviour, explicitly looking for widely applicable tendencies and the conditions under which such tendencies occur. While this is understandable when using theoretical frameworks that focus on and emphasise structures and structural interlinkages, in this chapter we have argued that the lived and subjective experience of actors is also important: system change is the result human action in some form and context.

Among the questions this raises is: (a) whether we should be looking for widely applicable 'rules' in *human* behaviour, i.e. maintaining a realist or positivist ontology; or (b) whether we want to continue with an 'outsiders' view of the sociotechnical systems in which actors are embedded. That is, whether there might be advantages to also considering these systems from a more social constructionist perspective, i.e. as perceived by the actors themselves. The latter could in principle provide a more flexible and empirically led view of how different parts of the system interact and the roles and perceptions of actors therein. This may facilitate deeper understandings of the actions (or inactions) of particular actors and agents within systems, adding nuance and informing as to the relations of power, dynamics and complexities within the parts of the system that they know best. In some contexts, then, it is precisely such system 'insider' knowledge that may be crucial for understanding the opportunities for – and the obstacles to – systems change. Indeed, the strong structuration (Stones 2005) perspective that we advocate here lends itself well to such multifaceted analysis.

General value of an extended structuration framework

Stones' (2005) framework situates the practices, beliefs and experiences of agents as central to social interaction and structure, and we suggest that this framework may also be useful for understanding some of the more social aspects of sociotechnical systems. This, in turn, may extend the explanatory capacity of the MLP (Upham *et al.* 2017). However, it requires a specification of the connections between the individual, the social and the sociotechnical: all agents are embedded in institutional contexts that influence their behaviour, and they are also influenced by other actors. At the meso level, institutions and organisations have particular institutional and organisational logics (Fuenfschilling and Truffer 2014). Over time, these logics can be changed through impetus from disruptive, external pressures. In the terminology of the MLP, such pressures may be from any of the three levels: the niche, the regime or the landscape (Geels and Schot 2007). To understand such change, Stones' (2005) strong structuration approach helps to link the four elements depicted in Figure 4.1: (1) agent psychology and agent 'conjunctural' knowledge; (2) the external system and structures in which agents operate; (3) agents' actions and practices that reproduce or challenge that structure or system; and (4) outcomes, i.e. change or stasis in that system and/or in agents' experiences.

In our empirical examples, we have also found glimpses of the differing national economic and innovation policy contexts, and we get a sense of how these contexts influence the experience of actors as professionals and, subsequently, how this influences their attitudes towards and perceptions of the technologies under scrutiny, hydrogen and HFCs. Activities in other countries may also influence the attitudes of these actors. Overall, the quotations above illustrate and underscore the value of understanding the role of individual actor perceptions in sociotechnical (structural) processes, but they also emphasise the value of taking into account how the actors experience these situations. This actor experience and knowledge is what Stones (2005) refers to as conjunctural knowledge, and it will both influence and be a part of the ongoing sociotechnical processes.

National differences: interviewee perceptions of the future for hydrogen and HFCs

As mentioned above, the research that we draw on here (Dütschke *et al.* 2017; Upham *et al.* 2017) includes quantified data that shows clear differences (as well as commonalities) across the countries studied. While we use some of this quantified data in the next chapter, here we continue with qualitative illustrations.

The interviewees' differences by nationality reflect a range of long-standing, interconnected factors. While country comparisons are always complicated, some cautious, overarching observations of sociotechnical expectations and wider perceptions by country are possible. The UK informants' perspective is shaped by a relatively individualised heating system, limited State involvement in utility provision and limited co-ordinating functions and powers in local and regional government, relative to other parts of Europe. Germany has a tradition of co-operative federalism that supports regional co-operation and a relatively decentralised fiscal system that supports investment in local infrastructure: regions and municipalities levy a higher proportion of tax revenue than in the UK relative to the national government (EUI 2009).

The UK and German actors have different experiences of what is possible in terms of infrastructure provision. Based on personal experience, the UK ministry actor 'knows' that injection of hydrogen to the gas grid and modification of gas boilers and hobs is more politically feasible than widespread installation of district heating. For the Spanish informant too, the emphasis on technical and economic limitations, as well as legal and regulatory complexities reflect the interviewee's personal experiences and shape his views of what is possible. In short, these actors are using their conjunctural knowledge to understand other actors (here, particularly, regulators, government and publics) and aspects of their own situation. This knowledge also informs their views on what is plausible and implausible as regards future action.

The German examples have a somewhat different tone: they explicitly and implicitly assume that HFC use for stationary applications *is* likely at some point in time. But these technologies are also manufactured by

German or Japanese firms, and this assumption is not premised on any advantages that HFCs may have relative to other options. There is confidence and assurance in the comments: the underlying assumption seems to be that in the long run HFCs for stationary applications will happen. One can only speculate as to the origins of this confidence, but being part of a billion-euro R&D programme focused on hydrogen must help. Conversely, the Spanish interviewees are working in a context of withdrawn subsidies for renewable energy, and the operating context for UK interviewees is somewhere in between. Thus, evidence suggests that expectations/anticipations among the informants regarding what is considered plausible do seem to be conditioned by the contexts in which they operate.

National similarities: interviewee perceptions of the future for hydrogen and HFCs

In this case study, most interviewees (of all the nationalities interviewed) believe that HFC heat and power technologies will remain in their niche (or niches) for the medium term at least. In the terminology of the MLP, a key reason for this is intra-niche competition. For example, electric vehicles are seen as strong competitors to hydrogen internal combustion and/or HFC vehicles in terms of mobility. Nonetheless, electric vehicles they are also seen as laying the necessary soft and hard infrastructure (electric power trains, associated workforce skills, etc.) required for HFCs as a subsequent and complementary technology.

However, in terms of meeting the need for heat and power, it is more difficult to see how HFCs would benefit from the installation of short-term, low carbon alternatives. Moreover, several interviewees emphasised the value of hydrogen as methanated or blended with natural gas for use directly in familiar, combustion-based technologies. In this regard, concerns were raised about the environmental logic and losses of multiple conversion phases (e.g. wind power to compressed electrolytic hydrogen, to fuel cell conversion to power or heat), and so the environmental benefits of HFCs need to be determined in relation to the alternatives in particular contexts. On the plus side is the perceived flexibility of hydrogen for crossing the regimes of heat, power and mobility: like electric power, hydrogen is an energy vector that can be used in many ways. Additionally, given its flexibility in terms of feedstock (hydrogen can be produced from many sources), and that the competences for its safe handling are principally found in the fossil gas sector (hence the skills base, mindset and basic technology for this already exists), hydrogen also has both strong potential and controversial connections with the fossil fuel sector (Upham *et al.* 2017). In other words, and particularly for heating use but also for transport, renewable hydrogen could in principle either replace fossil fuel use or, through blended use, continue the use of fossil natural gas.[10]

The importance of expectations for energy systems change

Overall, the above account proposes that subjective, situated experience informs actors' opinions or beliefs – in this case, of HFCs. This situated experience, in turn, may influence practice and thence the creation, reproduction or change in the structures and institutions that support these practices. Expectations, beliefs about the future and about what is possible, do play a role in realising technological futures (we pursue this theme in depth in Chapter 5). We know that there are many arenas and communities in which expectations relating to future technologies circulate and compete in more or less formal environments (Bakker 2012). We also know that at some point, those with positional authority can and do commit public and private sector resources that strengthen niche activities, potentially bringing them into the socio-cognitive 'regime' and expressed materially in the sociotechnical systems around us.

Importantly, while the expectations and actions of all types of actors in all parts of sociotechnical systems may prompt and facilitate change, they do so in the context of those existing technological infrastructures, production systems, socio-economic, policy and political contexts and so on, i.e. within the context of sociotechnical systems. As we have argued in this chapter, it is precisely this dynamic agency of actors within such sociotechnical systems that merits additional analytical attention in the sociotechnical transitions literature overall.

Conclusions

In this chapter, we have made a case for a strong structuration perspective for the analysis of agency–structure dynamics relating to energy systems and applications, analysed within sociotechnical frames. The intention has been to provide a theoretical account of agency for the sociotechnical transitions literature building on the MLP's premise of multiple levels of structuration in sociotechnical systems. Actors' beliefs, knowledge and experience are characterised as conjunctural knowledge (Giddens 1984). Based on the empirical data presented, we argue that R&D stakeholder beliefs of the prospects for HFCs for stationary applications are partly conditioned by their socio-economic and innovation policy context and that these beliefs, in turn, have consequences for sociotechnical change processes. That is, there are mutually influencing relationships among the actors and systems variables involved. The assertion of a mutual connection between the social and the technological is axiomatic in sociotechnical perspectives. Yet, it is still uncommon to find an emphasis on the role of individual beliefs in the sociotechnical literature.

Strong structuration draws attention to actors' conjunctural or situated knowledge as a potential influence on technological pathways and systems change. More generally, strong structuration supports the inclusion of psychological and subjective processes more centrally in accounts of niche–regime dynamics. Potentially, this flexibility embraces individualist psychological perspectives where behaviour is also seen as an outcome of individual-level factors (attitudes, norms, values, intentions, motivations, behaviours). Here, we have focused on sociotechnical beliefs, expectations and knowledge, but other possibilities for research enquiry could include identity and roles in motivating and facilitating structure change processes.

While it is challenging to evidence, document and generally research such processes and social systems dynamics, it is possible to work in an embedded way within changing contexts, documenting processes (Barley and Tolbert 1997) and using various indicators of change.[11] From a psychological perspective, other constructs (terms of analysis) are relevant and are the topic of this book. In short, for such research enquiry a broad spectrum of methods from the usual material of sociological and psychological analysis is likely to be beneficial. Such analysis is possible, then, even if in practice the research will capture only parts of the change processes.

For analysing energy systems-level processes, the sociotechnical transitions literature deploys the language of path dependency, path creation, logics, disruption, stabilisation, evolution, institutionalisation, emergence and so on. Yet all of these processes inevitably depend on agents as individuals and in groups, whose reflection, willingness and capacity to act are in part a function of their past and present situations. Overall, understanding agent-level characteristics and processes in relation to wider sociotechnical structural characteristics is a precondition for influencing such agent-level change processes, and these will provide part of the key to societal transitions to lower carbon energy systems.

Notes

1 The concept of habitus was developed by Bourdieu and adopted by Giddens. Initially Bourdieu (1977, p. 95) defined the idea thus:

> A system of lasting, transposable dispositions which, integrating past experiences, functions at every moment as a matrix of *perceptions, appreciations, and actions* and makes possible the achievement of infinitely diversified tasks, thanks to analogical transfers of schemes permitting the solution of similarly shaped problems.

Later Bourdieu modified this definition to the more general, "a system of acquired dispositions functioning on a practical level as categories of perception and assessment or as classificatory principles as well as being the organising principles of action" (Bourdieu 1990, p. 13, cited in Klapper 2011).

2 Values, in so far as they are slow-changing, can be conceptually located at the 'landscape' level; however, their expression is through individuals and their organising practices are at the regime level.

3 Socio-cognitive rules are immaterial, but they have material implications. They are the ways of thinking, including ideas and norms, that people use and follow and hence that structure regular social interaction, i.e. underpin social structure. In the MLP and more general sociotechnical terms, the regime is constituted of these non-material rules, while the sociotechnical system is the physical expression of this in terms of infrastructure, technology and society's relationships

with these (see Geels 2011). Despite these distinctions, however, in the socio-technical transitions literature, the regime is often treated as synonymous with the material sociotechnical system (Sorrell 2018).

4 This is a rough estimate only, and relates to heating, using the sources referred to below; there will presumably be more recent data available. Also note that mean (average) values obscure the variations between homes of different size and occupant profligacy. Graph 5i in Palmer and Cooper (2013) gives a 2009 value of CO_2 emissions per UK home (domestic space and water heating) of 5.5 annual metric tonnes. DECC (2014) cites the narrow 'commercial' potential of usable waste heat, defined as those projects which could provide a simple payback within two years of investment, as $1.1 MtCO_2/yr$ (5 TWh/yr), derived mostly from the use of heat exchangers to connect heat sources and heat sinks at the same site (1.1 million/5.5 = 200,000). Note that the use of a 'homes' equivalent for illustration does not mean that this waste heat could necessarily be economically transported to those homes.

5 http://hyacinthproject.eu/project/.

6 The German programme as a whole includes HFCs for all types of applications: stationary for heat and power; vehicular mobility and portable applications such as laptops, mobile phones; and small generators for e.g. mobile homes and temporary or emergency purposes.

7 The most fundamental rule for a commercial organisation is profit-making, while a public agency will (or should) seek the public interest. Individual actors will have different interpretations of these objectives and they will be more or less aware of others' interpretations.

8 Power to gas technology for grid balancing:

> Power-to-gas technology is a technological chain which converts the excess electricity into a gaseous fuel, such as hydrogen or methane … It can be used not only as an energy storage technology, but also as an instrument for balancing electric and gas networks.
>
> (Lewandowska-Bernat and Desideri 2018, p. 4570)

Balancing is usually understood temporally: ensuring that there is sufficient supply at the required times.

9 Hydrogen is difficult to store in a form useful for energy supply and hence research on hydrogen storage has been pursued for decades (Ren *et al.* 2017). The widely-held view among those active in this field is that the extensive utilisation of hydrogen as a fuel is being hindered by lack of effective hydrogen storage solutions (ibid.).

10 Hydrogen could also be produced from coal with carbon capture, but none of the interviewees referred to this.

11 Although it is not our objective to elaborate on these, such indicators include: changing discourse; changing institutional and organisational design, objectives and practice; social interaction patterns; investment decisions; historical perspectives and so on.

Bibliography

Aldous, D. 2014. "Understanding the complexity of the lived experiences of Foundation Degree sport lecturers within the context of Further Education". *Sport, Education and Society* 19(4):472–88. doi:10.1080/13573322.2012.674506.

Alonso, P. M., Hewitt, R., Pacheco, J. D., Bermejo, L. R., Jiménez, V. H., Guillén, J. V., Bressers, H. and de Boer, C. 2016. "Losing the roadmap: Renewable energy paralysis in Spain and its implications for the EU low carbon economy". *Renewable Energy* 89:680–94. doi:10.1016/j.renene.2015.12.004.

Ammermann, H., Hoff, P., Atanasiu, M., Aylor, J., Kaufmann, M. and Tisler, O. 2015. "Advancing Europe's energy systems: Stationary fuel cells in distributed generation". A study for the Fuel Cells and Hydrogen Joint Undertaking, European Union, Luxembourg. Available at: www.fch.europa.eu/sites/default/files/FCHJU_FuelCellDistributedGenerationCommercialization_0.pdf.

Archer, M. S. 1988. *Culture and Agency*. Cambridge: Cambridge University Press.

Archer, M. S. 1995. *Realist Social Theory: The Morphogenetic Approach*. Cambridge: Cambridge University Press.

Bakker, S., van Lente, H. and Meeus, M. T. H. 2012. "Credible expectations – The US Department of Energy's Hydrogen Program as enactor and selector of hydrogen technologies". *Technological Forecasting and Social Change* 79:1059–71. doi:10.1016/j.techfore.2011.09.007.

Barley, S. R. and Tolbert, P. S. 1997. "Institutionalization and structuration: Studying the links between action and institution". Industrial and Labor Relations School, Cornell University, NY. Available at: http://digitalcommons.ilr.cornell.edu/articles/130/.

Berger, R. 2015. "Now I see it, now I don't: Researcher's position and reflexivity in qualitative research". *Qualitative Research* 15(2):219–34. doi:10.1177/1468794112468475.

Bhaskar, R. 2014. *The Possibility of Naturalism: A Philosophical Critique of the Contemporary Human Sciences*. Abingdon, UK: Routledge.

Bodolica, V., Spraggon, M. and Tofan, G. 2016. "A structuration framework for bridging the macro–micro divide in health-care governance". *Health Expectations* 19(4):790–804. doi:10.1111/hex.12375.

Bourdieu, P. 1977. *Outline of a Theory of Practice*. Cambridge: Cambridge University Press.

Bourdieu, P. 1990. *In Other Words: Essays Towards a Reflexive Sociology*. Stanford, CA: Stanford University Press.

Casey, E. S. 2009. *Getting Back into Place*. Bloomington, IN: Indiana University Press.

CCC. 2017. "2017 report to Parliament – Meeting carbon budgets: Closing the policy gap". London: Committee on Climate Change.

Chan, Z. C., Fung, Y. and Chien, W. 2013. "Bracketing in phenomenology: Only undertaken in the data collection and analysis process?". *The Qualitative Report* 18(59):1–9. doi:10.1057/9781137326072.0007.

Coad, A., Jack, L. and Kholeif, A. O. R. 2015. "Structuration theory: Reflections on its further potential for management accounting research". *Qualitative Research in Accounting & Management* 12(2):153–71.

DBEIS. 2016. "Heat in buildings – The future of heat: Domestic buildings". Department for Business, Energy & Industrial Strategy, London, Available at: www.gov.uk/government/uploads/system/uploads/attachment_data/file/575299/Heat_in_Buildings_consultation_document_v1.pdf.

DCLG. 2017. "English Housing Survey 2015 to 2016: Headline report" (and accompanying tables). Department for Communities and Local Government, London. Available at: www.gov.uk/government/statistics/english-housing-survey-2015-to-2016-headline-report.

DECC. 2013. "The future of heating: Meeting the challenge". Department of Energy & Climate Change, London. Available at: https://assets.publishing.service.gov.uk/government/uploads/system/uploads/attachment_data/file/190149/16_04-DECC-The_Future_of_Heating_Accessible-10.pdf.

DECC. 2014. "The potential for recovering and using surplus heat from industry: Final report". Department of Energy & Climate Change, London. Available at; www.gov.uk/government/publications/the-potential-for-recovering-and-using-surplus-heat-from-industry.

Dütschke, E., Upham, P. and Schneider, U. 2017. "Report on results of the stakeholder survey". Deliverable 5.1., Centro Nacional del Hidrógeno (CNH2), Puertollano (Ciudad Real), Spain. Available at: http://hyacinthproject.eu/wp-content/uploads/2017/12/HYACINTH-D5_1-Report-on-results-of-the-stakeholders-survey_v02_DEF.pdf.

EUI. 2009. "Study on the division of powers between the European Union, the member states and regional and local authorities". European University Institute, Florence. Available at: https://cor.europa.eu/en/engage/studies/Documents/Study-Division-Powers/Study-Division-Powers-EN.pdf.

Fankhauser, S., Averchenkova, A. and Finnegan, J. 2018. "10 years of the UK Climate Change Act". Grantham Research Institute on Climate Change and the Environment, London.

Federal Ministry for Economic Affairs and Energy. 2015a. "Energy efficiency strategy for buildings methods for achieving a virtually climate-neutral building stock". Federal Ministry for Economic Affairs and Energy, Berlin. Available at: www.bmwi.de/Redaktion/EN/Publikationen/energy-efficiency-strategy-buildings.pdf?__blob=publicationFile&v=4.

Federal Ministry for Economic Affairs and Energy. 2015b. "Energiewende direkt, issue 09/2015". Federal Ministry for Economic Affairs and Energy, Berlin. Available at: www.bmwi-energiewende.de/EWD/Redaktion/EN/Newsletter/2015/09/Meldung/infografik-heizsysteme.html.

Feeney, O. and Pierce, B. 2016. "Strong structuration theory and accounting information: An empirical study". *Accounting, Auditing & Accountability Journal* 29(7):1152–76. doi:10.1108/AAAJ-07-2015-2130.

Fishbein, M. and Ajzen, I. 1975. *Belief, Attitude, Intention, and Behavior: An Introduction to Theory and Research*. Reading, MA: Addison-Wesley.

Fjellstedt, L. 2015. "Examining multidimensional resistance to organizational change: A strong structuration approach". PhD dissertation, The Graduate School of Education and Human Development, The George Washington University, ProQuest LLC, Ann Arbor, MI.

Fuenfschilling, L. and Truffer, B. 2014. "The structuration of socio-technical regimes – Conceptual foundations from institutional theory". *Research Policy* 43:772–91. doi:10.1016/j.respol.2013.10.010.

Garud, R. and Karnøe, P. 2001. "Path dependence as a process of mindful deviation". In: R. Garud and P. Karnøe (eds.), *Path Dependence and Creation*. Mahwah, NJ: Lawrence Earlbaum, pp. 1–40.

Gearing, R. E. 2004. "Bracketing in research: A typology". *Qualitative Health Research* 14(10):1429–52. doi:10.1177/1049732304270394.

Geels, F. W. 2002. "Technological transitions as evolutionary reconfiguration processes: A multi-level perspective and a case-study". *Research Policy* 31:1257–74.

Geels, F. W. 2011. "The multi-level perspective on sustainability transitions: Responses to seven criticisms". *Environmental Innovation and Societal Transitions* 1(1):24-40. doi:10.1016/j.eist.2011.02.002.

Geels, F. W. and Schot, J. W. 2007. "Typology of sociotechnical transition pathways". *Research Policy* 36:399–417.

Geels, F. W. and Schot, J. 2010. "Reflections: Process theory, causality and narrative explanation". In: J. Grin, J. Rotmans and J. Schot (eds.), *Transitions to Sustainable Development: New Directions in the Study of Long Term Transformative Change*. New York: Routledge, pp. 93–104.

Giddens, A. 1984. *The Constitution of Society: Outline of the Theory of Structuration*. Berkeley, CA: University of California Press.

Grin, J., Rotmans, J., Schot, J., Geels, F. and Loorbach, D. 2010. *Transitions to Sustainable Development: New Directions in the Study of Long Term Transformative Change*. New York: Routledge.

House of Commons Energy and Climate Change Committee. 2016. "2020 renewable heat and transport targets – Second report of session 2016–17". House of Commons, London. Available at: https://publications.parliament.uk/pa/cm201617/cmselect/cmenergy/173/173.pdf.

Klapper, R. 2011. "Social constructionism and constructivism in entrepreneurship research – A practical illustration of complementary perspectives". *Management & Avenir* 3(43):354–71.

Klapper, R. and Upham, P. 2015. "A model of small business and sustainable development: Values, value creation and the limits to codified knowledge tools". In: P. Kyro (ed.), *Handbook of Entrepreneurship and Sustainable Development Research*. Cheltenham, UK: Edward Elgar, pp. 275–98.

Lewandowska-Bernat, A. and Desideri, U. 2018. "Opportunities of power-to-gas technology in different energy systems architectures". *Applied Energy* 228:57–67. doi:10.1016/j.apenergy.2018.06.001.

Palmer, J. and Cooper, I. 2013. "UK housing energy fact file". Department of Energy and Climate Change. Available at: www.gov.uk/government/statistics/united-kingdom-housing-energy-fact-file-2013.

Parker, J. 2006. "Structuration's future?". *Journal of Critical Realism* 5:122–38. doi:10.1558/jocr.v5i1.122.

Ren, J., Musyoka, N. M., Langmi, H. W., Mathe, M. and Liao, S. 2017. "Current research trends and perspectives on materials-based hydrogen storage solutions: A critical review". *International Journal of Hydrogen Energy* 42:289–311. doi:10.1016/j.ijhydene.2016.11.195.

Schwandt, D. R. and Szabla, D. B. 2013. "Structuration theories and complex adaptive social systems: Inroads to describing human interaction dynamics". *Emergence: Complexity and Organization* 15(4):1–20.

Sorrell, S. 2018. "Explaining sociotechnical transitions: A critical realist perspective". *Research Policy* 47(7):1267–82. doi:10.1016/j.respol.2018.04.008.

Stones, R. 2005. *Structuration Theory*. New York: Palgrave Macmillan.

Svensson, O. and Nikoleris, A. 2018. "Structure reconsidered: Towards new foundations of explanatory transitions theory". *Research Policy* 47:462–73. doi:10.1016/j.respol.2017.12.007.

Tufford, L. and Newman, P. 2012. "Bracketing in qualitative research". *Qualitative Social Work* 11(1):80–96. doi:10.1177/1473325010368316.

Upham, P., Dütschke, E., Schneider, U., Oltra, C., Sala, R., Lores, M., Bögel, P. and Klapper, R. 2017. "Agency and structure in a sociotechnical transition: Hydrogen fuel cells, conjunctural knowledge and structuration in Europe". *Energy Research & Social Science* 37:163–74. doi:10.1016/j.erss.2017.09.040.

Weber, R. P. 1990. *Basic Content Analysis* (Second edn.). Newbury Park, CA: Sage Publications.

Part II
Case study applications

5 The psychology of expectations in sociotechnical systems

Introduction

This chapter illustrates how social psychology can add to existing, functional accounts of expectations in sociotechnical, specifically energy, futures. It is closely connected to the previous chapter, drawing in part on the same study. Again, the aim is not simply show the relevance of psychological theory and empirics for sociotechnical transitions perspectives of energy system change and stasis, but to show how these may be closely connected in ways that provide a fuller account of topics of interest to energy transitions processes.

In innovation studies and, more recently, sociotechnical transitions literature, what are described as 'expectations' play an important role (van Lente, 2012). 'Expectations' include all types of visions, scenarios, roadmaps and other depictions of the future, whether private or public, individual or collectively produced, formal or informal (ibid.). In general, expectations are viewed as directing attention (positive and negative) to future options, helping to direct and legitimise interest and investment (Borup *et al.* 2006; van Lente 2012). While innovation studies and the sociotechnical transitions literature tends to look at expectations through a functional lens, that is, at what expectations *do*, here we are more interested in their nature as a psychological phenomenon. We therefore define expectations as *beliefs about the future* and show how, interpreted in this way, expectations can be connected to one of the most well-known psychological models of behaviour – the theory of planned behaviour (TpB) – and hence can help to explain why individuals act positively on some expectations and not others.

Although the sociology of technological expectations literature primarily focuses on the social dynamics of expectations – such as how they circulate, compete and are involved in the 'hype' of changing levels of societal attention towards particular technologies (e.g. Alkemade and Suurs 2012) – the same literature also implicitly acknowledges their psychological dimensions. For example, those studying expectations have observed that one feature of foresight and road-mapping exercises, that is, formalised scenario processes usually involving a range of stakeholders, is the tendency to generate rather unoriginal visions of the future, perhaps in part due to those involved drawing on pre-existing cognitive repertoires of possibilities for that future (Jacobsson and Johnson 2000; Unruh 2000, cited in van Lente 2012). As these cognitive repertoires are both socially shaped and cognitively held, the question is how and why some such ideas and norms are internalised by individuals while others are not, and why some are given a lower degree of personal commitment than others. In both this and the next chapter we reflect upon these questions using different types of psychological theory. In this chapter, we refer to constructs that are common in cognitive, behavioural psychology, while in the next chapter we use the theory of social representations, which relates more directly to the acceptance of ideas.

We begin with a short overview of the literature on the sociology of expectations. We then outline three closely related psychological theories that are in essence refinements of the same proposition, namely, that behaviour follows from the belief that such action will have a desirable consequence for the actor: Vroom's (1964) expectancy theory; the theory of reasoned action (Fishbein 1967); and the theory of planned behaviour (Ajzen and Fishbein 1980). We use these theories to identify psychological constructs that connect the expectations of individuals (and, potentially, groups) to action. This constitutes more of a psychological perspective of what sociologists would describe as connecting agency to structure, and here more specifically to sociotechnical structures. To model these sociotechnical structures and processes we use the multi-level perspective (MLP) (Geels 2002), around which so much of the sociotechnical literature has coalesced (Sorrell 2018).

As illustrative case material, we use data from the same interviews with the European hydrogen innovation systems' actors referred to in Chapter 4, again specifically those involved with the use of hydrogen as a fuel and hydrogen fuel cell (HFC) electric vehicles. Thus, the empirical data for this study was collected as part of the same case, and analysed using the same methods described in Chapter 4.[1] In this chapter, however, we focus more specifically on *expectations* among the informants, and we show how the policy environment in which the informants are integrated interacts with those expectations as psychological constructs.

As caveats, our purpose is not to evidence this interaction in depth, but rather to show that we can look at sociotechnical expectations from long-standing psychological perspectives and to argue that this is likely to add to our understanding of the processes involved. Similarly, as the data available to us stops short of the level of detail required to connect specific behaviour to the psychological constructs that we identify, we have to be content here with setting out the rationale for a direction for further research. The chapter again draws upon on the study by Dütschke *et al.* (2017) and the interpretation of Upham *et al.* (2017). We begin with an overview of how technological expectations are usually considered.

The sociology of technological expectations

Within the literature on the sociology of shared expectations, such expectations are believed to help to co-ordinate action within and between organisations, acting as a 'constitutive force', particularly in the early stages of the social embedding (uptake and use) of an innovation. This is especially the case when different technology options (e.g. different types of energy supply etc.) are competing for investment and attention (Borup *et al.* 2006). It has been argued that shared expectations help to connect different groups within and between organisations (ibid.). In relation to one of the functions of innovation systems, this includes guiding the search for emergent technology selection[2] (Bergek *et al.* 2008; Hekkert and Negro 2009). Hence, facilitating shared expectations has become an important feature of "communication and interaction across institutional and epistemic borders" (Borup *et al.* 2006, p. 286), where the latter refers to the distinctions between different types of knowledge or knowledge domains, within or between organisations. Berkhout (2006) describes expectations as strategic "bids" for possible futures, highlighting their often competitive nature, especially at the early stages of sociotechnical development before major commitments to investment have been made.

Expectations are found everywhere. For those involved in formal, institutionalised visioning (typically well-resourced scenario production), Van Lente (2012) highlights some of the dilemmas posed by what he describes as a "sea" of informal expectations. Formal plans and visions for the future, such as roadmaps, scenarios and foresight activities take place not in a space empty of expectations but, conversely, in social environments in which a sea of expectations already exist, circulate and compete. If sociotechnical emergence from a niche requires the development of aligned and strong networks of actors, visions of the future that deviate substantially from each other may disrupt and eventually shape those networks.

With the above in mind, and thinking in terms of connections to psychological constructs, we suggest that expectations can be treated as *beliefs* about the future. Hence, to hold an expectation relating to, for example, hydrogen fuelled vehicles is to hold a belief about the future of those hydrogen fuelled vehicles. To view this expectation positively is to hold a positive *attitude* towards it – or the contrary. The *strength and the direction* of the attitude may be influenced by personal and social *norms*. The likelihood of acting in a way that assists the realisation of that expectation (belief) is considered the function of an *intention* to act.

Hydrogen fuel cell electric vehicles (HFCEVs) and hydrogen fuel cell applications in general are subject to what has been described as "arenas of expectations" (Bakker *et al.* 2011), where different actors promote and assess competing claims as to which technological option will succeed in the market (ibid., p. 55). As Bakker *et al.* (2011) emphasise, this process is collective, social and competitive at the same time, and it takes place in many different fora that include a wide variety of actors (industry actors, practitioners, policy makers, lobbyists, NGOs, end-users, news media, scientific conferences and journals, research funding agencies and so on). All of this activity taken together comprises an expectations dynamic that, assessed over time, can provide an indication of the future fortunes and prospects of particular technologies (Alkamade and Suurs 2012). HFCEV technologies have been described as being particularly subject to a 'promise–requirement cycle'. This means that performance-related promises (e.g. specific expectations regarding range, speed, safety and costs) with a specific time horizon (e.g. ten years) have become performance requirements as a condition for public and commercial investment (Van Lente 1993, 2000; Van Lente and Bakker 2010). In other words, product expectations that become product requirements are more likely to receive funding and investment and thus to become a reality (Upham *et al.* 2017).

Psychological theory and expectations

The range of theories that employ constructs directly relevant to psychological theories of expectations is rather small, although these theories are very commonly used. In particular, they are Vroom's (1964) expectancy theory and the related theories of reasoned action and planned behaviour (Fishbein 1967; Ajzen and Fishbein 1980).

Expectancy theory holds that the behavioural choice of an individual is a function of strength and direction of their attitudes (valence) and the perceived likelihood of a desired outcome following from that behaviour. Contemporary research using expectancy theory investigates the influences on valence and the anticipated outcomes of actions in, for example, organisational contexts. This could be the relation of employee work activities and associated organisational or employee goals (Chen *et al.* 2016).

The *theory of reasoned action* (Fishbein 1967) hypothesises that behavioural *intentions* are the best predictor of actual behaviour, and views those intentions as a function of the same core variables: attitudes to perceived action outcomes and the perceived likelihood of those outcomes being realised. The theory of reasoned action also takes into account social norms, e.g. societal beliefs about a behaviour.

The *theory of planned behaviour* adds the variable of perceived behavioural control. This includes factors that are both endogenous and exogenous to the individual, i.e. the locus of control as perceived by the individual may be seen as either internal or external to that person (Ajzen 2002). In the theory of planned behaviour, personal norms are regarded as differentiated from social norms (see Box 5.1).

Box 5.1 Theory of planned behaviour

The theory of planned behaviour (TpB) aims to predict how individual behaviours form, given the specifics of time and space, and to this end the theory emphasises the importance of motivation and abilities in determining behaviour. The theory identifies three main constructs that jointly contribute to determining the degree of control people have over their behaviours. These are: behavioural beliefs, normative beliefs and control beliefs. These overarching constructs and their subcategories refer both to societal norms and values *and* to the potential influence on/interpretations of these by the individual actors that ultimately 'do the behaviours'.

Behavioural beliefs refer to the attitude towards the act or behaviour; to the individual's assessment of how certain behaviours might ultimately become either a positive or a negative contribution in the life of that individual; to the consequences of actions. *Normative* beliefs focus on what is around the individual actors and, in this respect, to the normative expectations of others. For example, social groups and networks, cultural norms and group beliefs may influence individuals and their actions. Individuals may consider what peers and social network would think about doing something or about some action, and this again will influence the eventual behaviour. *Control* beliefs refer to how hard or how easy individuals believe it may be to do or display certain behaviours or agency, and thus these refer to beliefs about various factors that may either hinder or facilitate behaviours. Such assessments are inherently dependent on the specifics of time, place and context, and thus individuals can have very differing perceptions of their behavioural control given the particular situation.

Considering the impact of these constructs together, TpB predicts that a positive attitude towards the act or the behaviour, favourable social norms and a high level of perceived behavioural control are the best predictors for forming a behavioural intention, which in turn leads to a displayed behaviour or act. This is less like if any of the constructs prove less favourable (Ajzen 1991, 2006).

In Figure 5.1, key constructs are summarised and the model therein sets these in relation to the basic elements of the MLP of sociotechnical change (Geels 2002). As stated in Chapter 1, drawing on various literatures relating to technology development, the MLP posits that transitions come about through different types of interaction between processes at three levels: via *niche*-protected innovations gradually becoming more powerful; via *landscape*-level change that pressures the sociotechnical regime; and/or via destabilisation of the *regime*, enabling niche-innovations to gain their own momentum (Rip *et al.* 1995). At the micro level, niches are the location at which path-breaking innovations emerge – protected spaces that policies may passively or actively protect, nurture, empower or hinder (Smith and Raven 2012). At the macro level, the landscape is conceived of as an exogenous environment that is beyond the direct influence of niche and regime actors, including macroeconomics, deep cultural patterns and macro-political trends (Geels and Schot 2007) (see Box 5.2).

The model in Figure 5.1 depicts precisely the more individual level processes within the framework of the MLP, and it does so focussing on some of the most important variables considered in the theory of planned behaviour (Ajzen and Fishbein 1980). However, beyond simply depicting the psychological, social and sociotechnical processes in mutual

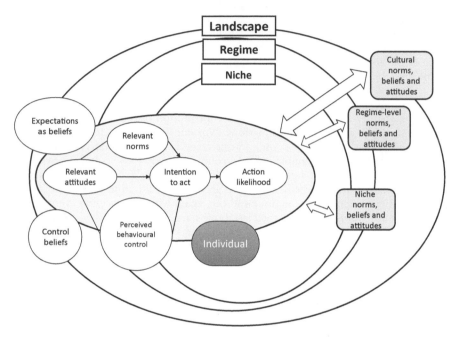

Figure 5.1 Basic components of individual psychology in relation to Geels' (2002) multi-level perspective of sociotechnical transitions.

Box 5.2 Exemplar reference to norms in the sociotechnical sustainability transitions literature

The multi-level perspective (MLP) recognises norms, beliefs and attitudes as operating at the niche, regime and landscape levels. For example, Elzen *et al.* (2011) discuss the role of normative contestation in transitions in relation to system innovation in pig husbandry. The authors describe how niche-level normative pressures from social movements for increased animal welfare in the Netherlands led to varying responses by some regime-level farmers, who then tested different pig husbandry designs.

In the MLP ontology overall (i.e. categorisation of phenomena at three levels), slow-changing social norms and values that are taken for granted are generally regarded as part of the 'landscape' level (Geels 2002). There is little intra-individual consideration of norms, values or attitudes within the transitions literature.

relationship, Figure 5.1 also implicitly connects the psychological aspects of expectations with the different levels of the *structuration* that underpins the niche, regime and landscape processes in the MLP (Geels and Schot 2007). Thus, the model also conveys psychological constructs and processes: (a) at levels beyond the individual, i.e. in different forms of collective organisation; (b) in relationships of mutual influence across these levels; and (c) in relationships of mutual influence with the systems that the individuals are part of. Importantly, in all of these processes *expectations* function as specific forms of belief. Overall, this connection of psychological and structuration processes helps to overcome the limitations of psychological perspectives that privilege a more individual level focus (see Batel *et al.* 2016), and at the same time it enriches the structure-focused accounts of change and resistance to change (Stones 2005).

Several further points can be made. First, following from the model, hope and fear are potential correlates of expectations. This is because expectations are often affectively engaged constructs, with emotional attributes. In this respect they differ from predictions per se (Castelfranchi 2005) – although one may also have hopes and fears about a prediction. Second, as humans are capable of abstractions, expectations involve cognitive representations of the future (ibid.). Third, and importantly, it follows from the model that not all expectations will lead to action. In this case, the main form of action would be efforts towards resource mobilisation, but a strong intention to act would also require sufficient strength and congruence among relevant attitudes, beliefs and norms of the actor(s) (see Castelfranchi 2005). In particular, one of the main implications of the TpB is that agents are less likely to act if they believe their actions will not have desirable consequences (Batel *et al.* 2016).

In the next sections we briefly describe the following: the policy context of our illustrative case; how the individual expectations of hydrogen and HFCEV stakeholders were elicited and analysed; and how these stakeholders' expectations illustrate the value of viewing expectations as future-oriented beliefs that interact with individuals' attitudes and norms. As we suggest, this is particularly important when seeking to understand the relationship between expectations and agency/action in sociotechnical contexts.

European policy context for HFCEVs

In the European Union, Directive 2014/94/EU on the deployment of alternative fuels infrastructure supports the development of low carbon transport. Here, alternative fuels are defined as: electricity, hydrogen, biofuels, synthetic and paraffinic fuels, natural gas (including biomethane), in gaseous form (compressed natural gas or CNG), liquefied form (liquefied natural gas or LNG) and liquefied petroleum gas (LPG) (Miguel *et al.* 2016). The Directive sets out minimum requirements for the development of alternative fuels infrastructure, and these minimum requirements include recharging points for electric vehicles and refuelling points for natural gas (LNG and CNG) and hydrogen, common technical specifications for these recharging and refuelling points, and user information requirements. Specifically, Article 5 of the Directive, "Hydrogen supply for transport", requires the circulation of hydrogen-powered motor vehicles within national transport networks with cross-border links by 31 December 2025. Deciding on the appropriate number of hydrogen refuelling stations to achieve this is left to individual member states (Miguel *et al.* 2016).

As mentioned in Chapter 4, in order to span high, medium and low HFCEV innovation activity in this study, five European countries are considered here: France, Germany, Slovenia, Spain and the UK. Among these countries, only Germany has a dedicated national hydrogen implementation plan. The German National Hydrogen and Fuel Cells Innovation Programme, managed by the National Organisation for Hydrogen and Fuel Cell Technology (NOW GmbH), is a public–private partnership spanning several ministries and regions with an initial, planned budget of 700 million euro from both government and industry (1.4 billion euro in total). A further initiative launched by NOW GmbH specifically for fuel cells transport application is H2Mobility, the first broad European plan to build a hydrogen refuelling station (HRS) network for HFCEVs (Miguel *et al.* 2016).

France has the strongest level of financial support from the State, particularly through the Hydrogen Mobility France initiative, a European financed project also involving four other countries. Partners in this project include the French government; energy companies; hydrogen and hydrogen refuelling producers; vehicle, fuel cell and electrolyser providers; research organisations; several regions; and EU and French associations. Key targets through to 2020 are to: create transnational corridors to Germany and Belgium for hydrogen vehicle use and refuelling; deploy HFCEV technology in clusters, i.e. so that specific regions are served; and to deploy extensive fleets of HFCEV vans, trucks and 15–20 hydrogen refuelling stations on selected road networks. The estimated funding needed until 2030 to implement the approximately 600 hydrogen refuelling stations and more than 800,000 HFCEVs necessary to achieve the longer-term goals of the programme is about 586 million euro (Miguel *et al.* 2016).

The UK, Spain and Slovenia all have project-based hydrogen and HFCEV activity, but there is less national programmatic support. In terms of project examples, in the UK, London had eight HFC buses under the Clean Hydrogen in European Cities (CHIC) project and Aberdeen had ten in 2016. This made Aberdeen's hydrogen bus fleet the largest in Europe at the time (Miguel *et al.* 2016). Through the multisector consortium UKH2Mobility in the UK, a national 2030 roadmap for HFCEVs has also been formulated.

All of the countries referred to here have forms of subsidy, tax incentives and related measures at the consumer level for low emission vehicles in general, though sometimes (as in Spain) only in certain regions. However, in terms of *national* private and/or public sector commitments to financial investment and infrastructure deployment relating to hydrogen applications, Germany's initiatives are the most substantial, while France has significant commitments too. It seems that where there are no nationally coordinated programmatic activities relating to hydrogen applications (as in Spain and Slovenia), initiatives such as hydrogen bus operations tend to cease as soon as the European funding expires (Miguel *et al.* 2016). While such project expiry applies to programme-funded activities too, individual project funding outside of programmes is usually on a more temporary basis. As we show via the presented data, such national policy differences do play a role in the psychology of the expectations of interviewees.

Individual stakeholder expectations

Below, we first provide an overview of some of the expectations that we identified among the informants and, second, we show how three psychological constructs – negative control beliefs, positive control beliefs and social norms – can help us to characterise and understand those expectations, both in and of themselves and in relation to anticipated action or agency. We now turn to an illustrative discussion of these expectations. Figures 5.2 and 5.3 show the expectations themes as identified among the interviewees; the figures illustrate the number of times each theme was mentioned by the countries studied.

The key expectations for hydrogen supply and use (independent of application) among the informants were generally positive, with market development expected in the relatively near term. Nonetheless, many of the interviewees did expect national differences in the way that supply and use of hydrogen will develop. Many informants expected the prospects and uses of an expanded hydrogen supply to be contingent on key decisions taken in the vehicle, heat and power sectors, but they also expected those decisions to be influenced by decisions taken outside of these sectors. In this regard, some informants mentioned the ongoing tightening of environmental legislation.

Many interviewees from across the five countries referred to Germany as most likely to be at the forefront of HFCEV developments due to its dedicated national hydrogen innovation programme and the substantial auto-manufacturing sector. French respondents were particularly confident of the prospects for hydrogen itself, perhaps because of a focus on (apparently successful) hydrogen trials in the professional transport sectors with light duty vehicle fleets fitted with a battery as a range extender. These offer the advantage, over the individual consumer sector, of the need for fewer refill points. That is, vehicles owned by professional fleets have shared and hence fewer home bases than is the case with individual owners – and fewer home bases means the need for fewer individual refill points.

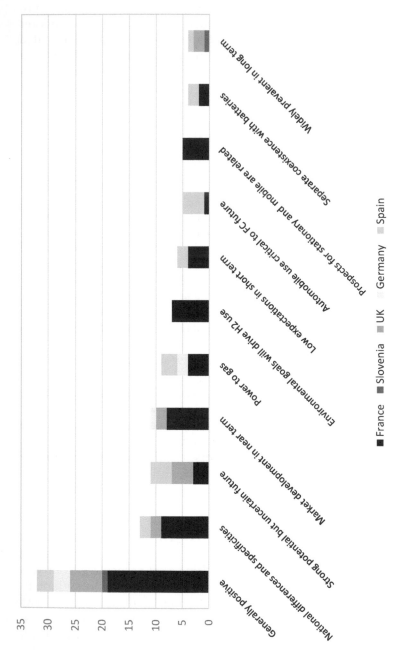

Figure 5.2 Hydrogen supply and use: expectations count.

Notes
1 H2 – hydrogen; FC – fuel cell.
2 Thirteen expectations expressed by 1–2 people per item have been omitted so that the figure legend remains legible.

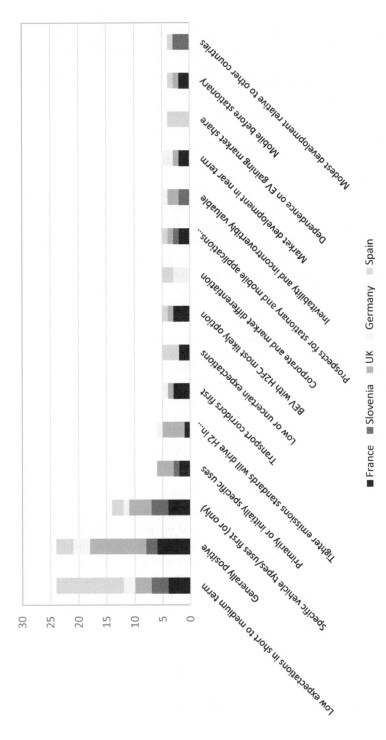

Figure 5.3 Hydrogen fuel cell electric vehicles: expectations count.

Notes

1 H2 – hydrogen; EV – electric vehicle; BEV – battery electric vehicle; H2FC – hydrogen fuel cell.

2 Thirteen expectations expressed by 1–2 people per item have been omitted so that the figure legend remains legible.

Overall, interviewees were very aware of their national and regional economic contexts and of the role and value of hydrogen supply not only for HFCEVs, but also the wider value of new regional economic clusters and connections with oil and gas industry expertise. The French interviewees were probably the most unequivocally optimistic about the prospects for hydrogen (see Figures 5.2 and 5.3).

For many interviewees, financial and cost-related matters in relation to HFCEVs were expected to be severely constraining on the further development of the technology. Generally, the financial instruments and forms of policy support intended to bring down the relative costs of hydrogen and HFCEVs (see e.g. OECD/IEA 2015) were seen as desirable possibilities, but they were also seen as unlikely in the short term. Hence, few expected HFCEVs to be deployed at any significant level in the short term. Rather, there were expectations of: HFCs and electric batteries being mutually supporting within individual vehicles; battery electric vehicles dominating in the short to medium and/or longer term; and hydrogen being blended with methane for compressed natural gas engines,[3] rather than used for HFCEVs. More positively for HFCEVs, some stakeholders expected the trend of electrification in transport to assist HFCEVs – which use electric powertrains – by making the latter and their supporting infrastructures more prevalent. In other words, more manufacturing capacity for suitable electric motor systems requires correspondingly skilled workforces, component supply chains, standards and, ideally, per unit lower costs. Hence, although battery electric vehicles were expected to compete with HFCEVs in the short to medium term, electrification pathways (consisting of all of the wide variety of actions necessary to transition from fossil-fuelled mobility to electric-powered mobility) would be established that would help HFCEVs in the long term (Kemp *et al.* 1998; Geels 2002; Geels *et al.* 2012; see Figures 5.2 and 5.3).

Psychological dimensions of expectations

In the next sections, we show how the basic psychological constructs of positive and negative control beliefs and social norms have the potential to help to explain how interviewees' experiences with hydrogen and HFCEVs influence their expectations of these and their propensity to act on those expectations, recommend action or expect action by others. This is not only likely to facilitate a richer characterisation of the position of agents than is usually found in literatures relating to technological expectations, but may also reveal some details about the nature of those expectations and their propensity for future material influence and change. We illustrate this with a number of interviewee quotations below; the selected quotations are categorised by type of psychological construct. As stated earlier, the data we have available can only take us so far in making the argument and it remains for further work to trace in empirical detail the connections we map out.

Negative control beliefs regarding hydrogen and HFCEVs

Many of the interviewees' responses relate to the control beliefs referred to in Box 5.1. Self- and response-efficacy are, respectively, the beliefs that acting/doing something will make little difference, either because of one's own limited capacity for action (self-efficacy) or because of the inconsequential nature of such action (response-efficacy). For the Spanish and Slovenian stakeholders in particular, negative control beliefs reflect limited government and/or commercial interest in hydrogen and HFCEVs at the national level.

The Spanish stakeholders referred to a lack of national governmental support for the sector:

> I do not have a crystal ball ... [but] what is clear is that Spain is not in the game. When we want to do it, it will be already done. Germans and others will have already done it ... It will happen as in the automotive sector, we are only good component manufacturers.
>
> (Spain, private sector employee)

Slovenian respondents also had limited expectations of the success of hydrogen and HFC technologies relative to other countries. For example:

> I estimate the role of this technology in Slovenia will be somewhat less than in the neighbouring countries (Austria, Germany etc.), but much stronger than in the south-eastern parts. Let's say we will be at about 50 per cent of the level achieved in Germany.
>
> (Slovenia, university employee)

However, German respondents too, had doubts about the *near-term* prospects for HFCEVs and the associated effectiveness of current R&D efforts:

> On ... the supplier side of the technology we do not have the necessary capacity that is needed. One major company is now consequently following this path, and they expect it to be relevant in the medium term, but many others are still playing on a low level. You cannot make billions from it at the moment. The time span is more towards 2030.
>
> (Germany, state research organisation employee)

In fact, this was the dominant message of all those interviewed, despite decades of investment in the sector to date. Apart from the currently very high cost of HFCEVs, many of the informants saw one of the key reasons for doubting the near-term effectiveness of R&D investments in HFCEVs and their associated technologies as the existence of alternatives that satisfy the same environmental and service provision objectives.

In the short to medium term, the EV [electric vehicle] is going to prevail. For road transport, I think [in the form of] liquefied natural gas. As an alternative to conventional fuels, I see more possibilities in EV because now there is more supply of vehicles and the cost of the recharging infrastructure is much more affordable. Manufacturers, due to the issue of CO_2 emissions, are positioning themselves in favour of electric vehicles. It is the most immediate.

(Spain, public sector employee)

Other informants expressed more confidence in non-fuel cell uses of hydrogen, however:

When we talk about hydrogen today, we always think about fuel cells. This is fine, but there are still very few HFCEVs, there is still no solution for heavy vehicles, etc. ... So then, we want to resume the tests done years ago to mix hydrogen and methane ... you can propel a bus running on natural gas with a mixture of 20 per cent hydrogen, you can reach even 25 to 26 per cent, and the rest natural gas.... The hydrogen field always seems to be in the skies, far away ... but with this process [mixing hydrogen with methane in natural gas vehicles], we are not talking about the future. We have a lot of faith in this strategy.... In the end, no one thinks that all cars will go with hydrogen, CNG [compressed natural gas] may be an alternative and, in the end, I think there may be room for hydrogen and natural gas.

(Spain, private sector employee)

Among all the informants, national governments were consistently seen as the most powerful actors in the sector, mostly due to their control over the national policy environments and hence fiscal measures, investment and various aspects of regulation.

A key question here is what level of importance will governments attach to moving towards low carbon technologies? The cheapest option is to continue with modern technology diesel cars etc. If we want to see decarbonisation of the transport sector it will be driven by governments which will see quicker roll out of low carb technology.

(UK, hydrogen partnership stakeholder)

I would see it [hydrogen technology] commercialising within ten years, that would be the hope, but that relies on what [governmental] policy support it secures.

(UK, public sector employee)

Positive control beliefs towards hydrogen and HFCs

Nonetheless, positive control beliefs were also expressed, particularly among the French informants. Implicit in many of the views expressed is the belief that ongoing investment is worthwhile, in that it will have a positive outcome – the latter is a core premise of the TpB and its antecedent theories:

> As this is a new technology, its costs are very high. Over time, these will decrease and it will become accessible to everyone.
>
> (France, public sector employee)

Sometimes, positive control beliefs were intertwined with positive beliefs about economic opportunities.

> I think there are opportunities for the countries that lead on these technologies. We see Germany going for hydrogen in a big way. Scotland has also done that. We've got clusters in Aberdeen, Orkney and Fife. We are placing ourselves well. There will be jobs created and opportunities for providing services and exporting as well. These are all opportunities that will come.
>
> (UK, private sector employee)

> I think the energy sector, the people who install the hydrogen equipment, and people who design it, designers, installers and manufacturers will have the most opportunities. It is the case that the electrolysers in our project are imported from Canada, so we need to grow a UK supply sector as far as we can. There are opportunities for UK companies to grow their businesses.
>
> (UK, non-profit organisation employee)

Additionally, positive control beliefs among interviewees from several countries were the result of – and associated with – the setting of realistic, achievable targets. For example:

> We have to start with the big vehicles: the buses.... Buses are big in consumption, intensity of use and they depend on one decision taker. Here is the issue. Then we will reach the passenger cars.
>
> (Spain, private sector employee)

Positive control beliefs are also associated with – and encouraged by – past experience:

> When I give a conference [paper] I have a presentation with one slide about the main car companies that in 2000 had planned to introduce

fuel cell vehicles. They had planned that in 2015 they would sell the first commercial hydrogen vehicles. And it has happened. It seems that the automotive industry has achieved what they had proposed. That's where there are more expectations, the introduction of hydrogen from the automotive sector; from there, hydrogen will begin to raise.

(Spain, foundation employee)

For several informants, positive control beliefs extended into a perception of the inevitability of the widespread use of hydrogen technologies in the long term:

Of course, the costs of the vehicles are still too high. Sure, there are a lot of problems to solve regarding the infrastructure (reliability, operating costs, the technology itself).... But in the long run there is no alternative, in my view.

(Germany, private sector employee)

Social norms, hydrogen and HFCEVs

Some interviewees referred to anticipation of positive attitudes and implicitly congruent social norms regarding the use of hydrogen and HFCEVs among the wider population, though this was seen as contingent on government support:

The consumer will positively welcome this type of technology if the governments encourage its use.

(France, private sector employee)

...there is European legislation around emissions, and the more that those emissions regulations are tightened and enforced the more it's going to make it more attractive for people to be operating zero emission vehicles, be this hydrogen or battery electric vehicles.

(UK, public sector employee)

Moreover, some of the visions among interviewees of changing social norms for use of these technologies where quite detailed:

I think there will have been a consistently growing intensity of regulations in city centres and a big drive to reduce emissions affecting urban populations. Initiatives for the development of hydrogen will have been in place, but by 2050 they will have been removed. Hydrogen will be a key part of the wider energy infrastructure which will not only include refuelling for different vehicles, but also a key part of storage and an appropriate use of local production of energy.

(UK, multisector partnership stakeholder)

Cars will probably have moved to a car share and car clubs basis, and people will have changed their attitude to car ownership. Any safety concerns will have been long addressed. Local governments will play a key role in the provision of hydrogen and shaping public attitudes supported by national governments and major companies.

(UK, multisector partnership stakeholder)

The results show how expectations of the future for HFCs also influence attitudes towards these technologies among the informants. In the data, control beliefs, and more overarching social norms regarding energy transitions, are shown to be of particular relevance. Both control beliefs and such social norms are strongly connected to sector developments at the system level; these connections are discussed in more detail below.

Connections to the multi-level perspective of sociotechnical change

In the case above, we have seen how individual-level processes interact with regime and landscape processes. The landscape as the domain of slow-moving change is expected by stakeholders to be supportive of hydrogen and HFCEVs in the long term, but less so in the short and medium term. As the data shows, this expectancy of change among informants is not simply in terms of the technology per se, but is also related to a change in attitudes towards and beliefs in factors such as climate change and the need for transitions to sustainability in general. Hence, pressures for an adequate response to climate change are expected to deepen over time, driving technological change in mobility beyond electric batteries and towards hydrogen, be this via fuel cell (and electric motor) or combustion (standard engine) technologies. Indeed, the view that this technological development was a 'natural' progression over time was expressed by some of the informants. Moreover, regulatory initiatives that have the aim of securing the quality of air locally were seen as another driver.

Such pressures may reflect shifting values or norms relating to environmental concerns, perhaps reframed in terms of societal well-being. In the psychological literature, norms tend to be defined as relating to what is understood as socially acceptable in terms of behaviour, and multiple behaviours may be consistent with a particular value (or attitude). Hence, at the landscape-level in the MLP, the widely shared value of protecting human health may be seen as consistent with a wide variety of social norms relating to sustainable mobility. Thus, for hydrogen and/or HFCs to benefit from the stronger commitment to mitigating climate change and improving local air quality that a number of interviewees expect, hydrogen technologies would also need to be perceived as a feasible and effective response to those climate change challenges.

Discussion

Using the same study discussed in Chapter 4, we have shown in this chapter how stakeholder expectations of a set of technologies can be understood in relation to some constructs that are widely used in variable-based environmental and sustainability-related psychology. In this way, we aim to contribute to the case for studying the connections between individual-level psychological processes and sociotechnical system-level dynamics. Here, those systems dynamics have been represented by the MLP (Geels 2002). We make a number of further observations below.

First, we note that while variable-based sustainable behaviour psychology (for want of a better term) by definition analyses in terms of separate psychological constructs and quantitative question scales, here we have used the same constructs to interpret and discuss qualitative data. Both approaches (quantitative and qualitative) are possible and, indeed, are likely to be complementary.

Second – as we reiterate throughout this book – we are not arguing for an exclusive focus on psychology, and this applies also to understanding the relationships between expectations and action. Some actors have more power than others and this has practical implications. Hence Stones' strong structuration referred to in Chapter 4 has often been combined with accounts of 'position-practice' (e.g. Coad and Glyptis 2014) – an approach that has been used to illustrate the influence of organisational role (position) in decision-making and outcomes. Moreover, actors may take into account and weigh up multiple, different expectations. The expectations of the informants interviewed here often referred to the strategic positioning and anticipated actions of other actors, something also highlighted in other studies of expectations (Budde *et al.* 2012).

More generally, the way in which technological expectations are socially situated is also emphasised by Berkhout (2006, p. 300): "... private expectations are to a large extent shaped by socially distributed rhetorics about the future, as well as by the inertias represented by material conditions". In our case, such socially distributed rhetorics can be seen as some of the shared expectations for the future of hydrogen and HFCEVs. These shared views and future expectations are, in a sense, illustrated in Figures 5.2 and 5.3. Here, each column represents a common view or expectation related to hydrogen and HFCEVs as shared across countries. As Berkhout also states (2006, p. 304):

> Regime members will align themselves to visions of the future that are aligned with their interests, and which they believe they have the resources to achieve (or which they believe they can convince other powerful actors to achieve with them).

Indeed, our interviewees often gave most weight to what was perceived as feasible in the present but, at the same time, they had an eye to the future. Indeed, several stakeholders did view a hydrogen focused future as inevitable – given time.

Conclusions

We argue that an individual-level psychological perspective, using even basic psychological constructs, can help to characterise salient aspects of actor expectations in relation to energy system change and stasis (and, indeed, other systems). This characterisation can contribute to a wider understanding of both individual-level expectations vis-à-vis particular technologies and the relation between these expectations and the more structural-level dynamics within sociotechnical systems. In the case examined above, different types of expectations of the future of hydrogen and HFCEV technologies can be understood as positive and negative control beliefs about the technologies – beliefs that may also include ideas of present and future social norms.

The beliefs held by the actors are the outcome of many influences, including national policy support as well as beliefs about the actions and nature of others, notably consumer-citizens. In this way, we have viewed sociotechnical expectations as involving both intra-individual (defined as psychological) and more inter-individual (defined as social) factors. By viewing expectations as beliefs and – in particular – beliefs that are associated with ideas of control and effectiveness, we add an extra dimension – a psychological dimension – to our understanding of the role of expectations in energy-related sociotechnical futures.

In the next chapter, we show the value of a different type of psychological approach, one which spans both psychological and sociological perspectives: Moscovici's (1984) theory of social representations.

Notes

1 For a detailed description of the methods used in this chapter, see Chapter 4.
2 An innovation system is an analytical construct for thinking about the relationships between actors and emerging technologies. The direction of search refers to the attention given to a wide range of factors, including future expectations and signals, that influence the propensity of a company to invest in the given area or product (Bergek *et al.* 2008).
3 For example, Nadaleti *et al.* (2017) model the environmental benefits of replacing diesel in city buses in Brazil with a blend of renewable hydrogen and biogas. The hydrogen would be produced through electrolysis when hydroelectric reservoirs are emptied for dam cleaning or water level control; and the biogas would be collected from city landfills. Using the biogas or hydrogen alone in combustion engines would be enough to replace the diesel fuel used by buses in the 27 states considered, and one could argue for a transition simply to waste biogas as being advisable. While it is still not economically feasible to store very large amounts of

hydrogen, a blend (e.g. 20 per cent hydrogen from waste hydro-electric energy) would extend the total fuel quantity and reduce hydrocarbon (HC), caron monoxide (CO) and CO_2 emissions from individual bus engines relative to biogas alone, but it would increase nitrogen oxide (NOx) emissions due to higher temperature combustion. All of this raises many more interesting issues for consideration, regarding priorities and trade-offs, but they are not our main focus here!

Bibliography

Ajzen, I. 1991. "The theory of planned behaviour". *Organizational Behavior and Human Decision Processes* 50:179–211. doi:10.1016/0749-5978(91)90020-T.

Ajzen, I. 2002. "Perceived behavioral control, self-efficacy, locus of control, and the theory of planned behavior". *Journal of Applied Social Psychology* 32(4):665–83.

Ajzen, I. 2006. "Behavioral interventions based on the theory of planned behavior". Available at: https://people.umass.edu/aizen/pdf/tpb.intervention.pdf.

Ajzen, I. and Fishbein, M. 1980. *Understanding Attitudes and Predicting Social Behavior*. Englewood Cliffs, NJ: Prentice-Hall.

Alkemade, F. and Suurs, R. A. A. 2012. "Patterns of expectations for emerging sustainable technologies". *Technological Forecasting and Social Change* 79(3):448–56.

Arfken, M. 2015. "Cognitive psychology". In: I. Parker (ed.), *Handbook of Critical Psychology*. Hove, UK: Routledge.

Bakker, S., Van Lente, H. and Meeus, M. T. H. 2011. "Arenas of expectations for hydrogen technologies". *Technological Forecasting and Social Change* 78(1):152–62.

Batel, S., Castro, P., Devine-Wright, P. and Howarth, C. 2016. "Developing a critical agenda to understand pro-environmental actions: Contributions from social representations and social practices theories". *Wiley Interdisciplinary Reviews: Climate Change* 7(5):727–45. doi:10.1002/wcc.417.

Bergek, A., Jacobsson, S., Carlsson, B., Lindmark, S. and Rickne, A. 2008. "Analyzing the functional dynamics of technological innovation systems: A scheme of analysis". *Research Policy* 37(3):407–29.

Berkhout, F. G. H. 2006. "Normative expectations in systems innovation". *Technology Analysis & Strategic Management* 18(3–4):299–311.

Borup, M., Brown, N., Konrad, K. and Van Lente, H. 2006. "The sociology of expectations in science and technology". *Technology Analysis & Strategic Management* 18(3–4):285–98.

Bows, A., with Anderson, K. and Upham, P. 2008. *Aviation and Climate Change: Lessons from European Policy*. London: Routledge.

Budde, B., Alkemade, F. and Weber, M. 2012. "Expectations as a key to understanding actor strategies in the field of fuel cell and hydrogen vehicles". *Technological Forecasting and Social Change* 79(6):1072–83.

Castelfranchi, C. 2005. "Mind as an anticipatory device: For a theory of expectations". In: M. De Gregorio, V. Di Maio, M. Frucci M. and C. Musio (eds.), *Brain, Vision, and Artificial Intelligence. BVAI 2005. Lecture Notes in Computer Science, 3704*. Berlin and Heidelberg: Springer, pp. 258–76.

Chen, L., Ellis, S. C. and Suresh, N. 2016. "A supplier development adoption framework using expectancy theory". *International Journal of Operations & Production Management* 36(5):592–615.

Coad, A. F. and Glyptis, L. G. 2014. "Structuration: A position–practice perspective and an illustrative study". *Critical Perspectives on Accounting* 25:142–61. doi:10.1016/j.cpa.2012.10.002.

Dütschke, E., Upham, P. and Schneider, U. 2017. "Report on results of the stakeholder survey". Deliverable 5.1., Centro Nacional del Hidrógeno (CNH2), Puertollano (Ciudad Real), Spain. Available at: http://hyacinthproject.eu/wp-content/uploads/2017/12/HYACINTH-D5_1-Report-on-results-of-the-stakeholders-survey_v02_DEF.pdf.

EEA. 2017. "Trends and projections in Europe 2017". European Environment Agency, Copenhagen. Available at: www.eea.europa.eu/themes/climate/trends-and-projections-in-europe/trends-and-projections-in-europe-2017 [accessed 26 May 2018].

Elzen, B., Geels, F. W., Leeuwis, C. and Van Mierlo, B. 2011. "Normative contestation in transitions 'in the making': Animal welfare concerns and system innovation in pig husbandry". *Research Policy* 40(2):263–75. doi:10.1016/j.respol.2010.09.018.

Fishbein, M. 1967. "Attitude and the prediction of behaviour". In: M. Fishbein, *Readings in Attitude Theory and Measurement*. New York: John Wiley & Sons, pp. 477–92.

Geels, F. W. 2002. "Technological transitions as evolutionary reconfiguration processes: A multi-level perspective and a case-study". *Research Policy* 31:1257–74.

Geels, F. W. and Schot, J. W. 2007. "Typology of sociotechnical transition pathways". *Research Policy* 36:399–417.

Geels, F., Kemp, R., Dudley, G. and Lyons, G. 2012. *Automobility in Transition? A Socio-Technical Analysis of Sustainable Transport*. New York: Routledge.

Hekkert, M. P. and Negro, S. O. 2009. "Functions of innovation systems as a framework to understand sustainable technological change: Empirical evidence for earlier claims". *Technological Forecasting and Social Change* 76(4):584–94.

Jacobsson, S. and Johnson, A. 2000. "The diffusion of renewable energy technology: An analytical framework and key issues for research". *Energy Policy* 28:625–40. doi:10.1016/S0301-4215(00)00041-0.

Kemp, R., Schot, J. and Hoogma, R. 1998. "Regime shifts to sustainability through processes of niche formation: The approach of strategic niche management". *Technology Analysis & Strategic Management* 10:175–98.

Klapper, R. 2005. "The Projet Entreprendre – an evaluation of an entrepreneurial project at a Grande Ecole in France". In: P. Kyrö and C. Carrier (eds.), *The Dynamics of Learning Entrepreneurship in a Cross-Cultural University Context*. Faculty of Education, University of Tampere, Finland, pp. 188–212.

Marcus, George E. 1995. *Technoscientific Imaginaries: Conversations, Profiles, and Memoirs*. Chicago, IL: University of Chicago Press.

Meinshausen, M. 2006. "What does a 2°C target mean for greenhouse gas concentrations? A brief analysis based on multi-gas emission pathways and several climate sensitivity uncertainty estimates". In: H. J. Schellenuber (ed.), *Avoiding Dangerous Climate Change*. New York: Cambridge University Press, pp. 265–79.

Miguel, E., Esteban, D., Rodríguez, L. and Auer, T. 2016. "European projects and policies". Deliverable 2.1 of WP2 Context Analysis, EU FP7 JFCU Project Hyacinth.

Moscovici, S. 1984. "The phenomenon of social representation". In: R. Farr and S. Moscovici (eds.), *Social Representations*. Cambridge: Cambridge University Press, pp. 3–70.

Nadaleti, W. C., Przybyla, G., Belli Filho, P., de Souza, S. N. M., Quadro, M. and Andreazza, R. 2017. "Methane–hydrogen fuel blends for SI engines in Brazilian public transport: Potential supply and environmental issues". *International Journal of Hydrogen Energy* 42:12615–28. doi:10.1016/j.ijhydene.2017.03.124.

OECD/IEA. 2015. "Technology roadmap: Hydrogen and fuel cells". Organisation for Economic Co-operation and Development/International Energy Agency, Paris. Available at: www.iea.org/publications/freepublications/publication/technology-roadmap-hydrogen-and-fuel-cells.html.

Ravasi, D. and Rindova, V. 2008. "Symbolic value creation". In: D. Barry and H. Hansen (eds.), *The SAGE Handbook of New Approaches in Management and Organization*. London: Sage Publications, pp. 270–84.

Rip, A. 1995. "Introduction of new technology: making use of recent insights from sociology and economics of technology". *Technology Analysis & Strategic Management* 7(4):417–32. doi:10.1080/09537329508524223.

Rip, A., Misa, T. J. and Schot, J. 1995. *Managing Technology in Society: The Approach of Constructive Technology Assessment*. London/New York: Pinter.

Schatzki, T. R., Cetina, K. and Savigny, E. V. (eds.) 2001. *The Practice Turn in Contemporary Theory*. London: Routledge.

Seamon, D. 2013. "Place attachment and phenomenology: The synergistic dynamism of place". In: L. C. Manzo and P. Devine-Wright (eds.), *Place Attachment: Advances in Theory, Methods and Applications*. London: Routledge, pp. 11–22.

Smith, A. and Raven, R. 2012. "What is protective space? Reconsidering niches in transitions to sustainability". *Research Policy* 41(6):1025–36.

Sorrell, S. 2018. "Explaining sociotechnical transitions: A critical realist perspective". *Research Policy* 47(7):1267–82. doi:10.1016/j.respol.2018.04.008.

Stones, R. 2005. *Structuration Theory*. Basingstoke, UK: Palgrave Macmillan.

Unruh, G. C. 2000. "Understanding carbon lock-in". *Energy Policy* 28:817–30.

Upham, P., Kivimaa, P. and Virkamäki, V. 2013. "Path dependency in transportation system policy: A comparison of Finland and the UK". *Journal of Transport Geography* 32:12–22.

Upham, P., Dütschke, E., Schneider, U., Oltra, C., Sala, R., Lores, M., Bögel, P. and Klapper, R. 2017. "Agency and structure in a sociotechnical transition: Hydrogen fuel cells, conjunctural knowledge and structuration in Europe". *Energy Research & Social Science* 37:163–74. doi:10.1016/j.erss.2017.09.040.

van der Hoeven, M. 2015. "Foreword". In: "Technology roadmap: Hydrogen and fuel cells". Organisation for Economic Co-operation and Development/International Energy Agency, Paris.

Van Lente, H. 1993. "Promising technology: The dynamics of expectations in technological developments". Doctoral thesis, University of Twente, Enschede, Netherlands.

Van Lente, H. 2000. "Forceful futures: From promise to requirement". In: N. Brown, B. Rappert and A. Webster (eds.), *Contested Futures. A Sociology of Prospective Technoscience*. London: Ashgate, pp. 43–64.

Van Lente, H. 2012. "Navigating foresight in a sea of expectations: Lessons from the sociology of expectations". *Technology Analysis & Strategic Management* 24(8):769–82.

Van Lente, H. and Bakker, S. 2010. "Competing expectations: The case of hydrogen storage technologies". *Technology Analysis & Strategic Management* 22(6):693–709.

Vroom, V. H. 1964. *Work and Motivation*. New York: John Wiley & Sons.

6 The role of social representations in sociotechnical transitions

Introduction

This chapter introduces a different socio-psychological approach to people's understanding of energy technology. Again, with reference to the multi-level perspective (MLP) as an exemplar model of sociotechnical change (Geels and Schot 2007), we set out a perspective on the interaction between niche, regime and landscape levels that deploys Moscovici's (1988) theory of *social representations*. From this perspective, we view actors in socio-energy systems as both producing and being affected by social representations via communication and interaction. We further suggest that social representations theory allows for a socio-psychological view of the three levels of the MLP. Thus, we view *social representations* held by actors at the three levels as interacting with each other, both consisting of and influencing material processes through the practices of the actors involved. This is not instead of material interactions, but supplementary to them, at an ideational level (in a 'battle of ideas'). The specific social psychological processes posited in social representations theory are discussed below.

Empirically, we examine these interactions through a case study of a technology that – at least in Europe – generates public controversy: fracking for shale gas. To this end, we draw on a comparative case study of social representations of shale gas in the UK, Germany and Poland by Upham *et al.* (2015). Shale gas exploitation has been described and represented in the national media of multiple European nations by a variety of actors from several perspectives, thus enabling a spread of social representations of fracking for shale gas to be observed. Currently, there is little work explicitly striving to connect social representations to sociotechnical processes, though there is work on social representations in relation to energy controversy (e.g. Batel and Devine-Wright 2014).

Social representations theory

Moscovici's (1988) social representations theory is a notable social psychological theory of the interrelationship of perception and social influence. The theory posits two key processes involved in understanding and evaluating changes in the social and physical environment: *anchoring* (categorising according to pre-existing cognitive frameworks, thus rendering the unfamiliar familiar); and *objectification* (translating the abstract into the concrete and tangible, usually involving mental imagery). Via these processes, new concepts acquire tangible and more 'real' qualities.

Moscovici (2000) argued that social representations have two functions. First, they conventionalise new concepts and give them a recognisable and common form, thus enhancing communication and coordination within a social group (ibid., p. 22). Second, social representations prescribe common ways of thinking about topics: "[social representations] are forced upon us, transmitted, and are the product of a whole sequence of elaborations and of changes which occur in the course of time and are the achievement of successive generations" (ibid., p. 24). In this way, Moscovici emphasises that social representations are not static. Instead, social representations are viewed as constantly changing, just as the communities through which they travel themselves change and take up other, new concepts, which in turn are anchored to older representations. In short, social representations are dynamic and cumulative processes, simultaneously ideational and cognitive (Upham *et al.* 2015).

Box 6.1 Social representations

When individuals or groups live together, interact and communicate, they need a system through which to communicate that captures common understandings. In a broad sense, these understandings are what Moscovici describes as "social representations". According to Moscovici, social representations are:

> systems of values, ideas and practices with a two-fold function; first, to establish an order which will enable individuals to orientate themselves in their material and social world and to master it; secondly, to enable communication to take place amongst members of a community by providing them with a code for social exchange and a code for naming and classifying unambiguously the various aspects of their world and their individual and group history.
>
> (Moscovici 1973)

Thus, as systems of beliefs, social representations have two main functions: "firstly, to allow a common ground for understanding reality which enables individuals and groups to guide themselves in relation to diverse social phenomena, and secondly, to provide consensual codes for communication and interaction in the social reality" (Moscovici 1973). Accordingly, communities typically share a set of social representations which organise the worldviews of community members and guides their interpretation of reality and their everyday practices. [These representations will also] share the conditions and constraints of access to power, both in terms of material resources and symbolic recognition.

(Campbell and Jovchelovitch 2000, p. 264)

"Negotiation, communication, and even resistance occur in the dialogical relationship between Self, Other and the objects of representations (e.g. ideas, theories, material objects)" (Marková 2003). These theoretical understandings point to the dynamic nature of representing (Moscovici 1998, 2008, cited in Medeiros 2017, pp. 17–21, 158).

Social representations constitute "forms of symbolic mediation firmly rooted in the public sphere" (Jovchelovitch 1996, p. 122). According to Moscovici (1988), these forms of symbolic mediation are shared in three different ways, depending on the relations between groups and individuals within society. First, certain representations are pervasive and shared across different groups. These *hegemonic representations* originate beyond the immediate context of meaning-making of specific social groups (Villas Bôas 2010). Thus, hegemonic representations are coercive and "embedded in culture and history and thus have a tendency towards stability" (Marková 2000, p. 455). Second, representations become *emancipated* to the extent to which social groups assimilate and transform their meanings according to the specific contexts of practice. For example, social representations of mental illness vary according to the political and pragmatic context of groups such as service users (Foster 2007) and mental health professionals (Morant 2006). Third, representations become *polemical* as they are the result of social struggle and resistance. These representations are not shared by all segments of society and "are often expressed in terms of a dialogue with an imaginary interlocutor [e.g. powerful social groups]" (Moscovici 1988, p. 222).

Bauer and Gaskell (1999) visualise the dynamic of social representation as a triangular relationship between: (a) the subjects, or carriers of the representation; (b) the object that is being represented; and (c) the "pragmatic context" of the group that holds the representation (ibid.). Thus, according to these authors, social representations theory can be illustrated visually as the interaction between the three points of this conceptual triangle (subject, object and context), with each point influencing the other two. Moreover, Bauer and Gaskell introduce an explicit time-axis in their visualisation, with the triangulation of subject, object and context changing over time. They represent this in a 'Toblerone' model of social representations (Bauer and Gaskell 1999), as illustrated in Figure 6.1.

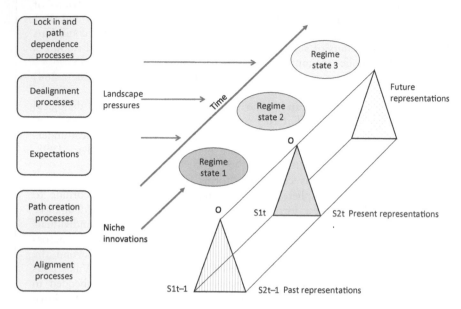

Figure 6.1 Processes of evolving social representation and sociotechnical change.

Source: follows Bauer and Gaskell 1999 and Upham *et al.* 2015.

Note

In the simplification posed by Bauer and Gaskell (1999), there are two subjects (individual perceivers) who carry a social representation, S1 and S2. O is the object of their representation and they are related in a project or domain represented by a triangle. This relationship has a past (t–1), a present (t) and a future. Maintaining the simplification, we take the domain in question to be the state of the regime for a particular sociotechnical system, itself subject to niche and landscape pressures, positing that the changing state of that regime is in part a function of changing and competing social representations. Niche representations are those that are marginal relative to dominant, regime-level representations. Landscape representations are long-lasting, relatively stable, representations.

Connecting social representations theory and the MLP

Both the MLP and social representations theory focus on different types and objects of change and involve different change processes. Yet, the two approaches can be readily connected. When Bauer and Gaskell (1999, 2008) refer to a triangular relationship between subject, object and context, this context (or conceptual domain) can be a sociotechnical system or parts thereof – as well as any other phenomena. The matter of theoretical connection then becomes one of investigating how social representations and related processes affect the processes of the MLP, including alternative pathways of transformation, reconfiguration, technological substitution, dealignment and realignment.

In Figure 6.1, we bring together graphically the basic concepts of the MLP and the dynamic processes of social representations over time, drawing on

Geels and Schot (2007), Bauer and Gaskell (1999) and Upham *et al.* (2015). As the figure illustrates, such connections may, in principle, be followed in detail longitudinally. Indeed, observations of the development of the full range of connections between evolving social representations and sociotechnical change requires a historical or longitudinal perspective, i.e. an extended period of time. The empirical cases used here depict the early stages of a quite contentious technology in the countries considered. As we will see, this enables observation of the spread, divergence and competition of social representations on a short timescale, all with differing implications for the societal embedding of a new (energy) technology.

As there is more than one subject (S1 and S2) who perceives object (O) in Figure 6.1, we are dealing with social representations, and not with the cognition of a single individual. The figure also shows multiple points in time (t-1, t and onwards into the future) as representations change. Moreover, Figure 6.1 juxtaposes the basic concepts of niche and landscape pressures, with regime states changing in response to landscape pressures. In this conception, social representations interplay in mutual relationship with the positions of political actors, institutions, corporations and other actors (Upham *et al.* 2015).

As observed previously, although not always referred to in studies making use of the MLP framework, the latter is ontologically founded on the cognitive, normative and regulative rules and institutions (very broadly interpreted) that are involved in co-ordinating human activities (Geels 2004; Geels and Schot 2007). Hence Geels and Schot draw on Bijker's (1995) idea of closure around a particular interpretation of a new technology, by which is usually meant agreement on the idea of what something is – for example, the idea that bicycles take a particular form.[1] Such closure "involves the build-up of a shared cognitive frame" (Geels and Schot 2007, p. 405) (in our terms, social representation) as an important aspect of sociotechnical transition processes. To this, Moscovici's theory adds the particular processes of *anchoring* and *objectification*, posited as underpinning shared perceptions and which, we suggest, contribute to or hinder sociotechnical change processes. Sociotechnical outcomes are the result of the interplay of all of the above and more, with social representations being one element.

As Geels and Schot (2007) emphasise, some ideas (and social representations are shared ideas) are more influential than others; this is a function of their social, political and material positions. Viewed in this way, representations that are less common remain niche-level representations, while representations that are cultural and relate to the most widely shared values form part of the slower-changing landscape. Moscovici's argument is that for new ideas to succeed, they need to become appropriated and integrated (anchored) into existing, shared socio-cognitive frames.

In the next sections, we illustrate some applications of social representations theory as a means of analysing the interconnections between ideas and energy technology acceptance – in this case fracking for shale gas – examined in the context of sociotechnical processes (see Upham *et al.* 2015).

Fracking for shale gas

Policy and practice context

In Europe, and particularly in Eastern Europe, exploratory drilling for shale gas has been undertaken by major oil and gas companies such as Total and Chevron, as well as by smaller operators (Williams *et al.* 2014), but the extent of the commercially viable shale gas resources in these regions remains to be seen. In many respects, fracking for shale gas can be described as a regime-level activity of the natural gas extraction industry (Westenhaus 2012). Fracking in the UK has previously involved mostly lesser known companies (Griffiths 2013), and in terms of the exploration and extraction of this natural resource, it is only more recently that one of the major oil and gas companies (Total) has become involved (BBC News 2014). In the UK, one of the large distribution firms, Centrica, has part-funded the exploratory activities of the firm Cuadrilla, formed in 2005 (Carrington 2014). In Poland, the possibility of shale gas exploitation has attracted both global companies (Total, Chevron and ExxonMobil) and Polish state-owned companies. Between 2007 and 2015, over 100 shale gas exploration licences were issued to over 30 companies. There has been some concentration of these licences; most belonged to Polish state-owned companies, while a few were in the hands of new players on the oil and gas market. The exploration licences covered almost 30 per cent of Poland's territory (Upham *et al.* 2015).

In Germany, a coalition government treaty has stated that unconventional gas exploration will not be undertaken, at least for the duration of the grand coalition government. This treaty followed a widespread furore when Chancellor Angela Merkel announced draft regulations permitting large-scale fracking in February 2013. This draft legislation was motivated by concerns over high energy costs, and it came from the Federal Department of Economics, led at the time by the pro-business Free Democrats. Following an influential 2012 report by the German Federal Environment Agency (Umweltbundesamt 2012), the coalition treaty against fracking was environmentally precautious. It stated that fracking is potentially a very high risk undertaking; that the use of environmentally toxic substances as part of fracking is rejected; and that a request for fracking approval can only be considered when any adverse change in water quality is categorically avoided (Deutschlands zukunft gestalten 2013).

Indeed, as Van de Graaf *et al.* (2018) highlight, European countries have developed very different responses to shale gas and fracking, from outright bans, to issuing permits and providing tax breaks. There are various possible reasons for these differences, including the prioritisation of energy security, economic competitiveness, political ideology, governance style and democratic tradition. Van de Graaf *et al.* (2018) find, however, that the only sufficient condition for restrictive regulation is the level of public concern. Other conditions only work in combination, while energy security and democratic tradition have no impact on the regulation of shale gas and fracking (ibid.).[2]

In the UK, attitudes to shale gas as captured in public opinion polls reflect considerable ambivalence and uncertainty. In June 2014, only about three-quarters of the general British public had heard of shale gas technology, and about half of those who knew of the technology neither opposed nor supported it. Of those who had heard of the technology at this date, support and opposition for it were equally split (DECC 2014). In Poland, approximately 59 per cent of the general public in areas surrounding shale gas exploitation supported these activities, while 78 per cent of the general public at the national level supported shale gas exploitation (CBOS 2013). German anti-fracking movements were stronger than their Polish equivalent, particularly in North Rhine-Westphalia and Lower Saxony (Lis 2014). These are also the most prospective shale gas regions in Germany. The German anti-fracking movements are also better organised and more far-reaching than the Polish movements. In Poland, the opposition seems mostly local, and there is seldom a request for a complete ban on fracking across the whole of Poland (Lis 2014; Upham *et al.* 2015).

Methods

The fracking case study presented here (Upham *et al.* 2015) documents social representations of this technology from print media (newspapers). The time period during which the data was collected was November 2011–June 2013 for the UK; January 2011–June 2013 for Germany; and June 2010–June 2013 for Poland. As such, the data represent a broad spread of news discourse on fracking and shale gas in the three countries during the period 2010–2013. The case study looked for and studied the differences and similarities in such representations between countries *and* between different types of newspapers. A spread of newspaper types reflected the possibility (and actuality) of different types of discourse. Hence, the newspaper sample encompassed both the more 'serious' press and titles that are known to be lighter in news content. The newspapers were sampled systematically so that the relative incidence of themes would be reflected. Sampling was used to select every *n*th article. In total, 303 newspaper items were manually coded from a total of 1,334 available, with the numbers of articles varying across the countries due to differing prevalence of relevant reporting and comment. The themes identified came from the data and coding process and were not established a priori.[3]

Results

United Kingdom

During 2012, *The Sun* (light news content) appeared unsure about the advantages and disadvantages of the fracking technology and where its own position on the matter lay. At this stage, the newspaper reported neutrally on the visit of a Nigerian campaigner to Ireland: "In my home country of Nigeria we have seen oil companies destroying our land ... Your country [Ireland] is beautiful – you must not let it be destroyed" (*The Sun*, 18 August 2012). By the end of 2012, however, *The Sun* had shifted to a strong use of humour and punning, combined with an emphasis on the potential benefits of shale gas not only at the national level, but in terms of reduced household bills. There was more of an emphasis on territoriality, with shades of patriotism, which are long-standing themes in the style of the newspaper.

Fracking was given the go-ahead by the UK government in mid-December 2012, but the likelihood of this development was anticipated and strengthened through officially commissioned reports such as that of Public Health England, which assessed the likely health impacts of shale gas fracking as low in October 2012 (PHE 2013). During the months prior to this, the potential economic value of the resource in the UK and the actual economic value of the resource in the United States appeared to coincide with a shift in both columnist and news articles, in both the broadsheets and *The Sun*, to a more concerted favouring of the exploitation of shale gas. This shift in representations over time is illustrated by *The Sun's* increasing reference to the household benefits of fracking and the newspaper's use of humour on the subject. For example:

> We need gas, Swampy ... so just frack off. Everyone's a winner. Except the nation's ecomentalists. They are furious. Because a cheap and plentiful supply of gas means society can keep on going. People can still drive cars and have patio heaters and take foreign holidays.
> (Jeremy Clarkson, *The Sun*, 15 December 2012)

Overall, the predominant themes on fracking in the UK during this more recent period are an appreciation and positive approval of the perceived benefits of fracked shale gas as a resource, with the broadsheets reporting the many commercial deals involved. Indeed, Total has since gone ahead with investment in UK fracking (BBC News 2014). In both UK newspaper types (broadsheets and tabloids), there was a strong theme of fracking-derived shale gas as a revolutionary energy resource, and this is portrayed through a variety of metaphors: as "tectonic plates shifting" (*The Sunday Times*, 24 February 2013); as the OPEC (Organization of the Petroleum Exporting Countries) "stranglehold" being broken by the shale "revolution" (*The Times* online, 17 January 2013); and as a "wake-up call that will change the world" (*The Independent* online, 14 December 2012). Shale gas fracking was also portrayed as a confounding of the thesis of peak oil:[4] "Bob Dudley, the chief executive [of BP], said: 'Conventional wisdom has been turned on its head. Fears of oil running out appear increasingly groundless'" (*The Independent* online, 17 January 2013).

Following UK government approval of fracking, a notable theme in the broadsheets during the period studied is that government support for fracking suggests a weakening of its position on climate change targets.

Germany

In Germany, the earliest newspaper articles on shale gas fracking, during the period studied, started to appear in 2011. The German press almost invariably described fracking as 'controversial' – even within the more positive or neutral articles. Thus, articles that outlined the views of proponents of fracking still referred to the technology as contested.

Overall, the coverage of fracking in German media has been negative. Fracking featured in the German news extensively during the latter part of the period surveyed (mid-2013). During this time, Germany was heading for a federal election (in September 2013), and several Landtag (regional) elections were also to be held during the period. Accordingly, politicians from all of the main parties aimed to clarify their positions on fracking as well as on other issues. Mindful of the generally negative attitude towards fracking technology among the German public, statements from the politicians rarely offered support for it (and especially so in the regions). Instead, positions ranged from complete rejection of the technology to very stringent demands of assurances on health and safety before fracking could be allowed. For example:

> following the Bavarian environment minister Marcel Huber ... [Christian Social Union], the SPD [Social Democratic Party] has also called for a fundamental prohibition of [shale] gas extraction using fracking, as long as dangers to the environment cannot be ruled out.
>
> (*Bild*, 1 March 2013)

Also notable in the German press are references to the relative 'cleanliness' of natural gas (fracked or otherwise and relative to the coal that the German energy system remains rather dependent on) in terms of CO_2 emissions: "[natural gas has] a reputation for being environmentally friendly..." (*Frankfurter Rundschau*, 30 May 2012).[5] In German newspapers, fracking has more generally been presented in terms of the benefits to the national economy, but less so for individual energy companies and more local economies. Questions of energy security featured mostly in discussions of the potential benefits of fracking for US energy security – and the resulting geopolitical changes – rather than with reference to Germany. Indeed, the geopolitical changes that fracking might drive were given substantial attention. This was sometimes framed positively, and sometimes in association with quite politicised opinions. For example, one *Bild* comment piece argued that as Saudi Arabia's influence wanes due to the West supplying its own fuels, it will be less likely to fund religious fundamentalist schools all around the world (*Bild*, 18 February 2013). More neutrally, others predicted that the geopolitical consequences of fracking in the United States could result in the reduction of the US military presence in the Middle East and elsewhere.

While geopolitics was a major theme of discussion in Germany, controversy over fracking has primarily centred on local environmental impacts, and often the key issue was contamination of groundwater supplies. The expression 'groundwater' ('Grundwasser': 10 references in the broadsheets, 11 in *Bild*) gradually shifted to 'drinking water' ('Trinkwasser': 14 references in the broadsheets, 25 in *Bild*) as the public debate on fracking became more heated, and as the newspaper articles – accordingly – shifted to a more negative stance. Generally, the terminology 'drinking water' may imply health risk connotations that appear less obvious when using the term 'groundwater'. The concern over drinking water contamination resulted in a somewhat idiosyncratic German furore over traditional German beer purity laws. This happened after the German Brewers Federation joined in the debate. They argued that if the quality of drinking water was jeopardised by fracking activities, they could potentially no longer guarantee uncontaminated beer, and this would ultimately endanger German 'cultural heritage' (*Bild*, 23 May 2013) (Upham *et al.* 2015).

Poland

The Polish case study shows how shale gas representations were changing over time, from primarily relating to economic and energy security issues, to being associated with specific issues of taxation and environmental regulations. One can also see how, through media representations, shale gas is cognitively integrated into the fossil fuel regime of energy production in Poland.

Between 2010 and 2012, two main themes were dominant in the newspaper *Gazeta Wyborcza*. Shale gas is described as geopolitically game changing[6] (a theme in 21 articles), and it is emphasised that it will enhance Poland's energy security (6 articles): "Shale gas is a great chance for Poland. Thanks to shale gas exploitation Poland may gain independence in energy policy" (*Gazeta Wyborcza*, 24 May 2013). Nonetheless, shale gas was also a controversial topic (11 articles), sometimes raising patriotic feelings (9 articles). For example, in September 2011 information was leaked to the press that Russian capital was behind companies with licences to explore shale gas in Poland, and that Russia may be trying to obstruct exploration processes in Poland (*Gazeta Wyborcza*, 23 September 2011). Subsequently, the government issued a statement that Polish regulations will prevent Russian companies from blocking shale gas exploration (*Gazeta Wyborcza*, 23 September 2011). This quite general fear of Russia's interference became more profound when, in mid-2012, Exxon Mobil announced that it was moving its exploration activities away from Poland. As the following quotation highlights, in some media this development was immediately connected to Russia's (supposed) interest in obstructing Poland's plans to exploit shale gas in the future: "Maybe it is not convenient for Exxon to invest in Poland because this is not what Gazprom with its stranglehold on our gas supplies wants" (*Gazeta Wyborcza*, 28 December 2010).

From 2010 until the end of 2013, *Gazeta Wyborcza* presented shale gas as having great economic importance for Poland (18 articles) as well as commercial value for companies (4 articles). In one article, the search for shale gas was even compared to the "gold rush", and the impact of the gold rush on the American economy was emphasised (*Gazeta Wyborcza*, 27 January 2010). In this context, however, the "gold rush" metaphor also referred to the interest of foreign companies in Poland's shale gas reserves: "in our country 'all the great ones of this world' – for example, Exxon Mobil, Chevron, ConocoPhillips, Marathon Oil Corp. – are looking for shale gas" (*Gazeta Wyborcza*, 27 January 2010). Moreover, the interest of these companies in Polish shale gas reserves brought about hopes for energy security: "they will earn money and we will gain energy security" (*Gazeta Wyborcza*, 27 January 2010).

In the newspaper *Dziennik Gazeta Prawna* shale gas was also framed as a unique chance for Poland to achieve independence from Russian gas supplies, with the added bonus of generating lower energy prices that could boost the Polish economy. In March 2012, a report on Polish shale gas reserves (carried out by the Polish Geological Institute) estimated the quantity of exploitable shale gas at about ten times *lower* than the earlier and very optimistic estimations made by the US Energy Information Agency (340–760 billion m^3 compared to 5,300 billion m^3). Thus, after 2012, new media representations of the shale gas technology emerged. These emphasised supply limits, the uncertain effects on fuel costs and concerns about environmental damage. The tabloid *FAKT* described this as "the end of the dream" (*FAKT*, 4 December 2012.).

FAKT has described environmental protestors as "hordes of eco-terrorists that came running to us to convince people that gas exploration is a tragedy which destroys the environment" (*FAKT*, 27 February 2012). In general, *FAKT* presents shale gas as controversial (18 articles) but, at the same time, manageable in terms of potential adverse effects of the technology (12 articles) and thus safe (10 articles). In the second half of 2012, *FAKT* published a series of ten educational articles in the form of questions and answers regarding shale gas and fracking (allegedly readers'

questions). These articles explained various aspects of shale gas exploration and fracking, its impact on the environment, risks and benefits. They
presented shale gas under headings such as: "we refute the myths", "the
good old fracking", "the law is protecting us" and "we are not afraid of
shale gas".[7] *FAKT* also underscored the potential of cheaper gas for households (*FAKT*, 13 September 2012 and 19 September 2012) (see Upham
et al. 2015).

Overall, representations can vary and compete, and what may be a form
of path creation for one actor can be seen as a form of path dependency
for another. These contrasts are illustrated in Table 6.1.

Discussion

Innovation studies and hence sociotechnical theorists recognise that "actors
use cognitive rules and schemas, some of which are shared" (Geels and Schot
2007, p. 415); that is, these rules, schemas, role relationships and normative
ties are used to interpret the world, create meaning and reach decisions
(ibid.). Here, in the context of fracking for shale gas, social representations
help bring to the point of enactment – including institutionalisation – imagined futures of prosperity and energy independence; or, conversely, of
helping to avoid damaged landscapes and a problematically warmer world,
as we fail to detach quickly enough from fossil fuel regimes.

It is in the sites and situations of sociotechnical controversy that we
perhaps best see the social dimensions of technological change: Doise
(1993, cited in Laszlo 1997) emphasises that social representations can distinguish – as well as bind – social groups together. This applies both within
and between countries. Moreover, social representations can be relatively
constant, or they can change over time. We see evidence of difference,
change and constancy in the media representations analysed here: in the
change in tone over time in *The Sun*; in the newspaper's usual style of
humorous punning and anticipation of its readership's concern with energy
bills; in the support for fracking offered in all the Polish newspapers examined; and in the consistent scepticism of the technology in the German
newspapers (Upham *et al.* 2015).

While there are some important commonalities across the countries,
notably issues of the need for domestic energy security and fracked shale
gas as changing the geopolitics of energy, there are also clear national
differences in the media discourses (and possibly also reporting protocols,
given the apparent sponsorship of articles in the Polish tabloid *FAKT*). As
Laszlo (1997) argues, *anchoring* as a concept in social representations
theory would hold limited analytic value if it was not culturally or socially
context specific. Hence, alternative anchors – i.e. alternative ways of
making new phenomena familiar – are almost inevitable and themselves
form part of the competitive environment in which different, partly
substitutable technologies are developed and promoted. Some of these

Table 6.1 Social representations of fracked shale gas (FSG) as dimensions of sociotechnical change processes in the media (examples from Poland, Germany and the UK)

Sociotechnical processes	Pro-shale gas representations	Anti-shale gas representations
Path creation	*Examples of FSG as a new energy source that is geopolitically revolutionary* *The Times* (online), 25 July 2012 and *Bild*, 19 April 2012: The United States may soon be self-sufficient in fuel, with profound economic, political and environmental consequences. *Dziennik Gazeta Prawna*, 15 November 2012: As a result of these changes, new coalitions will most probably be made on the geopolitical scene which will replace the old state of power relations.	*FSG as an unwanted variant in fossil fuel extraction* *Frankfurter Rundschau*, 21 May 2013: The risks posed by fracking chemicals exceed the benefits. *WNP*, 10 September 2013: Having respect for environmental goals and for the need to reduce greenhouse gases, especially CO_2, we decided to still bet on coal … Prime Minister Tusk added.
Path dependence	*FSG maintains existing ways of life* *The Sun*, 15 December 2012: A cheap and plentiful supply of gas means society can keep on going. People can still drive cars and have patio heaters and take foreign holidays. *WNP*, 2 September 2013: Lower costs of energy will increase competitiveness of the industries and may attract new investments, especially in the situation when the previous investors' attractor – the lower labour costs – is slowly disappearing.	*FSG entrenches fossil fuels* *Die Welt*, 21 April 2011: The shale gas boom could undermine investments in renewable energies.
Expectations and visions	*FSG will support national and commercial prosperity* *Tagesspiegel*, 31 May 2013: … with the controversial extraction method of fracking, energy prices will be lowered considerably. *WNP*, 2 September 2013: Shale gas can be a chance for re-industrialisation in Poland. *Gazeta Wyborcza*, 27 May 2011: Abundance of domestic gas is also a chance for modern chemical industries, petrochemical industries and for the production of equipment for shale gas extraction.	*Investor concerns about FSG* *The Independent* (online), 15 June 2012: Oil and gas explorers come under pressure to clamp down on controversial extraction process. *Dziennik Gazeta Prawna*, 22 May 2013: In order to succeed we need legal regulation favourable for risky gas extraction investment – and we do not have such, a representative of the gas industry notes.

continued

Table 6.1 Continued

Sociotechnical processes	Pro-shale gas representations	Anti-shale gas representations
Alignment – regime-level	*FSG fits with the existing sociotechnical system (positive)* *The Times* (online), 24 February 2013: We are looking at power shortages in two or three years, and the first new nuclear plant is at least a decade away. Fracking is a godsend. *Tagesspiegel*, 13 February 2012: Natural gas will become the oil of the twenty-first century. *Dziennik Gazeta Prawna*, 15 November 2012: Oil and coal dominate global energy consumption. Natural gas has a large share of the market as well – the future belongs to gas.	*FSG fits with the existing sociotechnical system (negative/mixed)* *Frankfurter Rundschau*, 2 February 2013: It is debatable whether it will be worth it for the gas industry, because every piece of regulation will drive the costs higher.
Dealignment – landscape-level	*FSG-induced geopolitical change viewed positively* *The Sun*, 21 September 2012: The prospect of a self-sufficient Britain is causing alarm in gas-rich Russia, where President Vladimir Putin assumed he had Western Europe at his mercy. *Tagesspiegel*, 22 February 2013: Should something horrible happen in the Middle East, then I can easily imagine that a US president would say: 'I don't care. We have enough energy.' *Dziennik Gazeta Prawna*, 29 April 2013: Such developments may mean that Russia, the main player on the global conventional gas market, will start losing around 1 per cent of PKB [GDP] each year.	*FSG-related geopolitics as posing dilemmas* *Bild*, 29 May 2013: Germans would import Russian gas without worrying about environmental impacts in Russia … I would call this hypocrisy, Oettinger [EC Commissioner for Energy] said.

Note

The comments included in the table are quotations (or translated quotations) from the newspaper sources cited.

anchors cross well between countries and cultures and others do not. The German 'drinking water' anchor will cross national boundaries to some extent, but the connection to beer production is a special and nationally specific link. Thus, anchoring is always related to the local context, to situated practices. Social representations of these practices, which can lead to different objectifications of risks, both contribute to and reflect different policy results (Upham *et al.* 2015).

In Poland, shale gas is aligned with a valued fossil fuel regime through representations of economic prosperity and energy independence from its powerful neighbour Russia, and related economic and geopolitical visions constitute the main framing for interpreting shale gas exploration activities. Thus, when Exxon Mobil announces its withdrawal from the country, this is also aligned with the fear of Russian influence on Polish energy supplies. Interestingly, in the Polish articles examined, shale gas was never presented as a threat to the Polish coal sector. A relative scarcity of environmental themes in Polish media discourse on shale gas underscores the conclusion that the finding of shale gas does not herald any major technological shift, but rather a path-dependent development in the Polish energy sector (Upham *et al.* 2015).

While the UK newspaper *The Sun* eventually treats shale gas as just another form of heating fuel, albeit one that the UK potentially owns, the UK broadsheet group and the Polish *Dziennik Gazeta Prawna* more typically treat shale gas as corporate news of commercial and economic significance. Most of the UK broadsheet articles on shale gas appear in the business sections of the newspapers. By contrast, the most prevalent themes of articles on shale gas in the German *Die Welt* focus on controversy and environmental concerns. Opposition to fracking in Germany during the period when the data was collected was also substantial, and it mostly related to concerns about potential local environmental impacts. In each of the three countries studied, the international, geopolitical implications of fracking and shale gas were frequently discussed and repeated.

Conclusions

Social representations are internalised, socially shared ideas with which people make sense of the world. They have many functions, including the capacity to propose, defend and justify particular perspectives. Here we have drawn on Upham *et al.* (2015) to highlight and illustrate sociotechnical processes relating to energy, in which social representations arguably play a role; we have used media (newspaper) based representations of fracking in Poland, Germany and the UK as illustrative examples. We have shown that the European countries considered have recent but differing histories of fracking debate and practice. Using these, we connected sociotechnical and social representations theory and allocated social representations to particular processes in sociotechnical systems change. We thus helped to reveal some of the psychological aspects of the dynamics involved in sociotechnical change, specifically in relation to energy supply (Upham *et al*. 2015).

Different technologies involve particular social representations of specific interactions of niches, regimes and particular aspects of the landscape (in MLP terms). In some ways, fracking for shale gas is unusual because fracking is a practice as well as a set of interrelated technologies. Moreover, shale gas as a commercial resource is not novel at all: the first commercial natural gas extracted from shale wells in the United States was reportedly dug in 1821 in Fredonia, Chautauqua County (NYSDEC 2007). Hence, natural gas from shale wells, its supply and distribution infrastructure and the companies and institutions involved in its delivery were established long ago. In many ways, shale gas fracking is thus the product of an incumbent sector. Nonetheless, the technology of hydraulic fracturing and the scale of exploitation of this resource are new globally, and especially new to Europe and its publics (Upham *et al*. 2015).

Social representations offer a psychologically informed and focused account of agency and of differential social embedding of phenomena – and it is often *new* phenomena that are studied. In this chapter, by connecting Moscovici's processes of anchoring (and to a lesser extent objectification) to sociotechnical change processes, we have introduced another way of thinking about these connections. Moreover, the simplicity of the social representations approach – and the way in which it can make use of historical representations as data – readily lend it to further application via extended historical as well as contemporaneous case studies. Social representations theory can be critiqued for its blend of ideation, discourse and psychology (Mckinlay and Potter 1987), but arguably Moscovici's thesis reflects the way in which these are closely related in practice, and the way in which ideas about technology can be as powerful as technology itself.

Notes

1 Strictly speaking, a bicycle has two wheels, but variants have one to four wheels, can have covered or uncovered sections, be with or without trailers, child or second person seats at the front or rear, come with electrical assistance or not, etc. Different countries have different ideas and norms of what is desirable and possible in terms of bicycles. The ideas and the materiality reinforce each other. Without a safe space to cycle, cycling itself feels dangerous and is more likely (we would suggest) to take a restricted form.

2 Van de Graaf *et al*. (2018) used fuzzy qualitative comparative analysis (QCA), a case-oriented method for systematically comparing cases in terms of three to eight conditions. Each country is scored for its performance or status against each condition. Obviously, this scoring process is both critical to the method and also debatable.

3 Full details of the method and more detailed results are given in Upham *et al*. (2015): here we focus on showing how the documented representations have implications for societal embedding of the technology (in this case, fracking for shale gas), and also how representations reflect the changing debate about this over time.

4 The idea that the global maximum production capacity of conventional sources of crude oil has already been reached or soon will be, with ongoing decline in production thereafter. Note that in social representations terms, fracked shale gas is being 'anchored' to oil here.

5 In 2017, 35.6 per cent of German electric power production was from two forms of coal: hard coal and lignite. Carbon intensity is a measure of the quantity of carbon or CO_2 emitted per unit of power production. Fuels vary in their carbon intensity. Exact values vary by extraction and conversion technology, as well as measurement. Nonetheless, natural gas is in general regarded as having approximately half of the carbon intensity of coal. The carbon intensity of European Union power production varies considerably across countries, corresponding to their generating supply mix. In units of CO_2 per kWh, the European Environment Agency (EEA 2017) gives values of, for example, 275.9 for the EU28; 34.8 for France (reflecting its high level of nuclear power); 424.9 for Germany; and 388.8 for the UK (2014 values). European Union power production is becoming less carbon intensive, albeit slowly: a reduction from $431 gCO_2/kWh$ in 1990 to $275.9 gCO_2/kWh$ in 2014, i.e. a reduction of 36 per cent or -1.44 per cent per year if averaged (EEA 2017). Industrial offshoring to countries with higher power system emission intensities has reduced the global net effectiveness of reduced CO_2 emissions from power production: measured at a global level, CO_2 emissions are growing in parallel with electricity demand. Following a 98 per cent increase in electricity generation worldwide between 1990 and 2014, CO_2 emissions from electricity generation also increased by 87 per cent, largely due to a lack of improvement in global average carbon intensity, which improved only marginally from $540 gCO_2/kWh$ in 1990 to $510 gCO_2/kWh$ in 2014 (Goh *et al.* 2018).

6 The geopolitics of fracked shale gas are much discussed but take us too far from our topic to enter into in depth here. The main immediate geopolitical issues include the suppression of global oil and gas prices; the effect of this on politically stable and unstable states that are dependent on fossil fuel derived revenue; and the implications of changing patterns of international energy dependence and independence.

7 This series of articles appeared to be sponsored by a gas company, as the name of one such company was repeated in most of the articles.

Bibliography

Arthur, B. 1989. "Competing technologies, increasing returns, and lock-in by historical events". *The Economic Journal* 99(394):116–31.

Batel, S. and Devine-Wright, P. 2014. "Towards a better understanding of people's responses to renewable energy technologies: Insights from social representations theory". *Public Understanding of Science* 24:1–15. doi:10.1177/0963662513 514165.

Bauer, M. and Gaskell, G. 1999. "Towards a paradigm for research on social representations". *Journal for the Theory of Social Behaviour* 29(2):163–86.

Bauer, M. and Gaskell, G. 2008. "Social representations theory: A progressive research programme for social psychology". *Journal for the Theory of Social Behaviour* 38(4):335–53.

BBC News. 2014. "French oil giant Total to invest in UK shale gas". *BBC News* [online], 14 January 2014. Available at: www.bbc.co.uk/news/uk-25695813 [accessed 21 January 2014].

Bijker, W. E. 1995. *Of Bicycles, Bakelites and Bulbs: Towards a Theory of Sociotechnical Change*. Cambridge, MA: MIT Press.

Callon, M. 1991. "Techno-economic networks and irreversibility". In: J. Law (ed.), *A Sociology of Monsters: Essays on Power, Technology and Domination*. London and New York: Routledge, pp. 132–61.

Campbell, C. and Jovchelovitch, S. 2000. "Health, community and development: Towards a social psychology of participation". *Journal of Community and Applied Social Psychology* 10(4):255–70.

Carrington, D. 2014. "Emails reveal UK helped shale gas industry manage fracking opposition". *Guardian* [online], 17 January 2014. Available at: www.theguardian.com/environment/2014/jan/17/emails-uk-shale-gas-fracking-opposition.

CBOS. 2013. "Społeczny stosunek do gazu łupkowego" [Social attitudes towards shale gas]. Public opinion research conducted in May 2013. Fundacja Centrum Badania Opinii Społecznej, Warsaw. Available at: www.cbos.pl/SPISKOM.POL/2013/K_076_13.PDF [accessed 1 October 2013].

DECC. 2014. "DECC public attitudes tracker – Wave 10". Department of Energy and Climate Change, London. Available at: www.gov.uk/government/uploads/system/uploads/attachment_data/file/342426/Wave_10_findings_of_DECC_Public_Attitudes_Tracker_FINAL.pdf.

Deutschlands zukunft gestalten. 2013. "Koalitionsvertrag, zwischen CDU, CSU, SPD". Available at: www.bundesregierung.de/breg-de/themen/koalitionsvertrag-zwischen-cdu-csu-und-spd-195906 [accessed 14 January 2014].

Doise, W. 1993. "Debating social representations". In: G. M. Breakwell and D. Canter (eds.), *Empirical Approaches to Social Representations*. Oxford: Oxford University Press.

EEA. 2017. *Overview of electricity production and use in Europe* [online], European Environment Agency. Available at: www.eea.europa.eu/data-and-maps/indicators/overview-of-the-electricity-production-2/assessment [accessed 7 December 2018].

Foster, J. L. H. 2007. *Journeys Through Mental Illness: Clients' Experiences and Understandings of Mental Distress*. Basingstoke, UK: Macmillan Press.

Garud, R., Kumaraswamy, A. and Karnøe, P. 2010. "Path dependence or path creation?". *Journal of Management Studies* 47:760–74.

Geels, F. W. 2002. "Technological transitions as evolutionary reconfiguration processes: A multi-level perspective and a case-study". *Research Policy* 31(8–9):1257–74.

Geels, F. W. 2004. "From sectoral systems of innovation to socio-technical systems: Insights about dynamics and change from sociology and institutional theory". *Research Policy* 33:897–920.

Geels, F. W. 2014. "Regime resistance against low-carbon transitions: Introducing politics and power into the multi-level perspective". *Theory, Culture & Society* 31(5):21-40. doi:10.1177/0263276414531627.

Geels, F. W. and Schot, J. 2007. "Typology of sociotechnical transition pathways". *Research Policy* 36(3):399–417.

Geels, F. W., Kern, F., Fuchs, G., Hinderer, N., Kungl, G., Mylan, J., Neukirch, M. and Wassermann, S. 2016. "The enactment of socio-technical transition pathways: A reformulated typology and a comparative multi-level analysis of the German and UK low-carbon electricity transitions (1990–2014)". *Research Policy* 45:896–913. doi:10.1016/j.respol.2016.01.015.

Goh, T., Ang, B. W., Su, B. and Wang, H. 2018. "Drivers of stagnating global carbon intensity of electricity and the way forward". *Energy Policy* 113:149–56. doi:10.1016/j.enpol.2017.10.058.

Griffiths, 2013. "Shale gas rush: The fracking companies hoping to strike it rich". *Guardian* [online], 12 March 2013. Available at: www.theguardian.com/environment/2013/mar/12/shale-gas-rush-fracking-companies [accessed 21 January 2014].

Howarth, C. A. 2006. "Social representation is not a quiet thing: Exploring the critical potential of social representations theory". *British Journal of Social Psychology* 45(1):65–86.

Jackman, M. and Sterczyńska, S. 2013. "Gaz z łupków w oczach mieszkańców, samorządów, koncesjonariuszy i instytucji województwa pomorskiego" [Shale gas in the perception of the inhabitants, local authorities, license holders and institutions from the Pomorskie Region]. *Przegląd Geologiczny* 61(1):381–5.

Jovchelovitch S. 1996. "In defence of representations". *Journal for the Theory of Social Behaviour* 26:121–35. doi:10.1111/j.1468-5914.1996.tb00525.x.

Laszlo, J. 1997. "Narrative organisation of social representations". *Papers on Social Representations* 6(2):155–72.

Levidow, L. and Upham, P. 2017. "Linking the multi-level perspective with social representations theory: Gasifiers as a niche innovation reinforcing the energy-from-waste (EfW) regime". *Technological Forecasting and Social Change* 120:1–13. doi:10.1016/j.techfore.2017.03.028.

Lis, A. 2014. "Public controversies over shale gas in Europe – Germany". News service of the Polish Geological Survey, Warsaw.

Mckinlay, A. and Potter, J. 1987. "Social representations: A conceptual critique". *Journal for the Theory of Social Behaviour* 17:471–87.

Marková, I. 2000. "Amedee or how to get rid of it: Social representations from a dialogical perspective". *Culture & Psychology* 6(4):419–60. doi:10.1177/1354067X0064002.

Marková, I. 2003. "Constitution of the self: Intersubjectivity and dialogicality". *Culture & Psychology* 9(3):249–59. doi:10.1177/1354067X030093006.

Meadowcroft, J. 2009. "What about the politics? Sustainable development, transition management, and long term energy transitions". *Policy Sciences* 42:323–40.

Medeiros, B. 2017. *Ageing well in the community: social representations of well-being promotion later in life*. Doctoral thesis, St. Edmund's College, University of Cambridge.

Montgomery, C. T. and Smith, M. B. 2010. "Hydraulic fracturing: History of an enduring technology". *Journal of Petroleum Technology*, December 2010. Available at: www.ourenergypolicy.org/wp-content/uploads/2013/07/Hydraulic.pdf [accessed 30 December 2014].

Morant, N. 2006. "Social representations and professional knowledge: The representation of mental illness among mental health practitioners". *British Journal of Social Psychology* (British Psychological Society) 45(Pt 4):817–38.

Moscovici, S. 1973. "Foreword". In: C. Herzlich (ed.), *Health and Illness: A Social Psychological Analysis*. New York: Academic Press.

Moscovici, S. 1988. "Notes towards a description of social representations". *European Journal of Social Psychology* 18:211–50.

Moscovici, S. 1998. "The history and actuality of social representations". In: U. Flick (ed.), *The Psychology of the Social*. Cambridge: Cambridge University Press, pp. 209–47.

Moscovici, S. 2000. *Social Representations*. Cambridge: Polity Press.

Moscovici, S. 2008. *Psychoanalysis: Its Image and Its Public*. Cambridge: Polity Press.

NYSDEC. 2007. "New York's oil and natural gas history – A long story, but not the final chapter". New York State Department of Environmental Conservation. Available at: www.dec.ny.gov/docs/materials_minerals_pdf/nyserda2.pdf [accessed 14 January 2014].

PHE. 2013. "Review of the potential public health impacts of exposures to chemical and radioactive pollutants as a result of the shale gas extraction process". Public Health England, London. Available at: https://assets.publishing.service. gov.uk/government/uploads/system/uploads/attachment_data/file/332837/PHE-CRCE-009_3-7-14.pdf [accessed 27 November 2018].

Phillips, N., Lawrence, T. and Hardy, C. 2004. "Discourse and institutions". *Academy of Management Review* 29(4):635–52.

Polish Shale. 2013. "Mieszkańcy Lubelszczyzny popierają wydobywanie gazu łupkowego" [Inhabitants of Lubelskie support shale gas exploration], *Prawo.pl* [online], 4 March 2013. Available at: www.prawo.pl/biznes/mieszkancy-lubelszczyzny-popieraja-wydobywanie-gazu-lupkowego,158662.html [accessed 5 December 2018].

Smith, A. 2007. "Translating sustainabilities between green niches and socio-technical regimes". *Technology Analysis & Strategic Management* 19(4):427–50.

"Środowiskowe aspekty poszukiwań i produkcji gazu ziemnego łupkowego i ropy naftowej łupkowej" [online resource] 2011. Ministerstwo Środowiska oraz Państwowy Instytut Geologiczny- Państwowy Instytut Badawczy, Warsaw. Available at: www.mos.gov.pl/g2/big/2011_08/d048c148f02734c384a119266ba159b8. pdf [accessed 27 March 2014].

Umweltbundesamt. 2012. "Umweltauswirkungen von Fracking bei der Aufsuchung und Gewinnung von Erdgas aus unkonventionellen Lagerstätten". Available at: www.umweltbundesamt.de/sites/default/files/medien/461/publikationen/4346.pdf.

Upham, P., Lis, A., Riesch, H. and Stankiewicz, P. 2015. "Addressing social representations in socio-technical transitions with the case of shale gas". *Environmental Innovation and Societal Transitions* 16:120–41. doi:10.1016/j.eist.2015.01.004.

Van de Graaf, T., Haesebrouck, T. and Debaere, P. 2018. "Fractured politics? The comparative regulation of shale gas in Europe". *Journal of European Public Policy* 25:1276–93. doi:10.1080/13501763.2017.1301985.

Villas Bôas, L. P. S. 2010. "Uma abordagem da historicidade das representações sociais". *Cadernos de Pesquisa* 40(140):379–405.

Westenhaus, B. 2012. "New fracking technology to bring huge supplies of oil and gas to the market". *Oilprice.com*, 16 January 2012. Available at: http://oilprice. com/Energy/Natural-Gas/New-Fracking-Technology-To-Bring-Huge-Supplies-Of-Oil-And-Gas-To-The-Market.html [accessed 22 January 2014].

Whitmarsh L. 2012. "How useful is the multi-level perspective for transport and sustainability research?" *Journal of Transport Geography* 24:483–7.

Williams, S., Amiel, G. and Scheck. J. "How a giant Kazakh oil project went awry", *The Wall Street Journal*, 31 March 2014.

7 The role of values in grassroots innovations

Introduction

In this chapter, we illustrate the role of *values* in sociotechnical transitions, again using a psychological perspective, drawing on Martin and Upham (2016)[1] for both the empirical case and theory. The case relates to the online free reuse groups Freecycle (which operates internationally) and Freegle (UK only), which promote and facilitate the reuse of 'stuff' (typically, unwanted household items), and are characterised here as *grassroots innovations*. While the case is not directly related to energy supply and demand, the connection of course lies in the embodied energy of goods and the corresponding benefits of product life extension. Product life extension for sustainability reasons requires balancing the improving energy and other performance of new products against the various environmental (and other) costs of new manufacture and purchase. In general, extending the use life of relatively materially dense, currently short lifespan and low energy consumption products (such as laptops) is likely to be advantageous in terms of total life-cycle impact, relative to making a new purchase (Bakker *et al.* 2014). However, this advantage also holds if the higher energy consuming product to be reused (such as a refrigerator) was bought relatively recently – i.e. relative to an older version of the product, the new fridge could be reused for longer (ibid.). One could say that reuse and increased product lifetime extension of consumer goods is thus even more justifiable in energy terms than it used to be.

Turning to the terms used here, grassroots innovations are bottom-up initiatives (Seyfang and Smith 2007) that are typically not for profit and which may offer a challenge to the dominant regime. They are by definition niche-level activities. The question posed by Martin and Upham (2016) is whether values might play a constraining role in the scale-up of such grassroots innovations. That is, whether the values of the participants involved in grassroots innovations differ so much from those of the dominant values in the regime, that this difference plays a restrictive role for grassroots activities. The case study builds upon the results of a large scale survey of the users of these online free reuse groups and, drawing upon this data, Martin and Upham show how such 'grassroots innovation' networks reflect and mobilise the values of the citizens involved. In this process, the authors develop a conceptual model of those processes as spanning two scales of analysis: (1) the individual scale, exploring which values are held by people who participate in such initiatives; and (2) the collective scale – which values are held by the wider society that they seek to mobilise. The study brings together Schwarz's social psychological theory of cross-cultural values (Schwartz 1992, 2006; Schwartz *et al.* 2012) and the sociological theory on the collective enactment of values (Chen *et al.* 2013).

The Freecycle and Freegle reuse groups examined in the case study have, in total, millions of members across the world. Their purpose is to help people to freely and directly give unwanted 'stuff' to others in their local area for reuse – rather than just throwing these things away and sending them to their local authority waste management system (Freecycle 2014; Freegle 2014). As such, online free reuse groups can be viewed as enabling a form of collaborative consumption (Botsman and Rogers 2011), holding the potential to both help reduce consumption of new resources and the production of waste, by extending product lifetimes. As mentioned in the opening paragraph, reuse has close, if somewhat complicated, connections to energy consumption.

Grassroots innovations

The sociotechnical sustainability transitions literature has tended to focus on the potential of technological innovations and the market economy to drive the transition to a sustainable society (Markard *et al.* 2012; Smith *et al.* 2010). However, there is also an interest in *civil society* as an overlooked context from which 'grassroots social innovations' with the potential to contribute to this transition may also emerge (Seyfang and Smith 2007). Seyfang and Smith (2007, p. 585) develop a model of grassroots innovations "to describe innovative networks of activists and organisations that lead bottom-up solutions for sustainable development; solutions that respond to the local situation and the interests and values of the communities involved".

In the next sections, we outline some of the research on grassroots innovations that are relevant for this study, and we describe the theories of values that have been applied. The case study results are then presented, and finally the empirical and conceptual implications for the diffusion of grassroots innovations in sociotechnical transitions are discussed.

Theoretical perspective

Grassroots innovation research has mostly focused on the dynamics of international and national networks of social economy and civil society actors (Vergragt *et al.* 2014). Such networks of grassroots innovation connect societal experiments in the form of community-based initiatives grounded in specific local contexts, and explore alternative configurations of urban production and consumption systems (Heiskanen *et al.* 2015). Studies of grassroots innovation include explorations of the promises and perils of community energy systems (Hargreaves *et al.* 2013a), cohousing provision (Boyer 2014), community currencies (Seyfang and Longhurst 2013), local food production systems (Kirwan *et al.* 2013) and democratic innovation systems (Smith *et al.* 2014).

From small beginnings around 2007 onwards (Seyfang and Smith 2007; Smith *et al.* 2010; Markard *et al.* 2012), the literature using the term 'grassroots innovation' in the explicit context of sociotechnical transitions has grown (Mokter 2018), though not all of the present literature is concerned with sociotechnical transitions theory. The early literature, in particular, discussed the somewhat limited applicability of aspects of theory that was originally developed to explain the dynamics of technological and market-driven innovation (e.g. Seyfang and Longhurst 2013). While such studies have focussed on community activities driven by radical (deep green) values, and the role of values in driving grassroots innovations is frequently acknowledged (Mokter 2018), the study presented below remains among the few that consider values centrally, particularly in relation to sociotechnical systems processes.[2]

Values are a contested but widely and variously used concept in the social sciences. The literatures that relate to values are substantial, spanning large areas of social psychology and sociology. Hitlin and Piliavin (2004) and Dietz *et al.* (2005) offer at least a starting point for an overview. Values are theorised to be held and enacted at multiple scales and Martin and Upham (2016) differentiate between individual (Schwartz 1992), collective (Chen *et al.* 2013) and cultural values. Martin and Upham (2016) integrate theory, enabling the exploration of the values of participants in societal experiments generally, along with the ways in which collective activities, such as societal experiments, are shaped by and seek to shape values.

From a behavioural psychological perspective, in particular, individual values are usually theorised as cognitive constructs with motivational implications. Schwartz and Bilsky (1987, p. 551) identify five core features of values: "values are (a) concepts or beliefs, (b) about desirable end states or behaviors, (c) that transcend specific situations, (d) guide selection or evaluation of behavior and events, and (e) are ordered by relative importance". Schwartz (1992) developed a prominent theory of individual values, which has been applied in hundreds of research studies (Schwartz *et al.* 2012). This theory identifies ten basic values (see Table 7.1 and Box 7.1), and Schwartz argues that they are grounded in universal human requirements for survival and existence, including our biological needs and the need for social coordination (Schwartz 1992). Conceptually, these ten values are represented as a circular motivational continuum (see Figure 7.1)

Table 7.1 Conceptual definitions of Schwartz's ten basic values and their relation to the overall motivational goals

Basic value	Definitions of basic values according to their motivational goals	Overall value	Values unacted by participants in free reuse groups
Self-direction	Independent thought and action.	Openness to Change	• forming connections with members of a local community and meeting new people when giving away or receiving an item (Nelson et al. 2007; Foden 2012);
Stimulation	Choosing, creating, exploring excitement, novelty and challenge in life.		• engaging in alternative forms of waste disposal, consumption and/or charitable giving practice (Guillard and Bucchia 2012b);
Hedonism	Pleasure and sensuous gratification for oneself.		• freely choosing who to give an item to (self-direction).
Achievement	Personal success through demonstrating competence according to social standards.	Self-enhancement	
Power	Social status and prestige, control or dominance over people and resources.		
Security	Safety, harmony and stability of society, of relationships and of self.	Conservation	• engaging in an act that extends the stewardship and life of an item, and hence in a basic sense conserves resources and exercises frugality;
Conformity	Restraint of actions, inclinations and impulses likely to upset or harm others and violate social expectations or norms.		• performing the practice of thrift (Foden 2012) – although this may be a necessity for some members, it may be a traditional practice for others;
Tradition	Respect, commitment and acceptance of the customs and ideas that traditional culture or religion provides.		• engaging with one's local community – a traditional activity that predates the atomised communities of capitalist society (see Putnam 2000).
Benevolence	Preservation and enhancement of the welfare of people with whom one is in frequent personal contact.	Self-transcendence	• reducing one's environmental impact by extending the lifespan of an item (Foden 2012);
Universalism	Understanding, appreciation, tolerance and protection for the welfare of all people and for nature.		• helping someone in need to obtain an item that could improve their quality of life (Nelson and Rademacher 2009; Groomes and Seyfang 2012; Guillard and Bucchia 2012b); • supporting one's local community (Nelson et al. 2007).

Source: Schwartz 1992; Schwartz et al. 2012.

Note
Some explanatory examples of how participants in free reuse groups can and do enact these values through the groups are included in the final column.

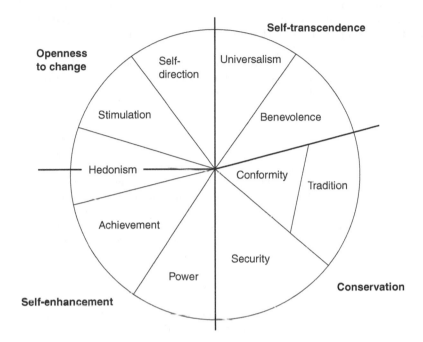

Figure 7.1 Theoretical model of relations between Schwartz's ten basic values.

Note
Original figure by Schwartz (1992) and Schwartz *et al.* (2012); this version reproduced under Creative Commons licence NonCommercial-NoDerivatives 4.0 International (CC BY-NC-ND 4.0).

where the distinction between adjacent values is blurred (Schwartz 1992; Schwartz *et al.* 2012). The proximity or distance between a given pair of values suggests the degree of compatibility or potential conflict between them.[3] Each of the ten values are connected to one of four broader, more abstract values, these being: openness to change, conservation, self-transcendence and self-enhancement (see Figure 7.1). Two scales for measuring the importance an individual places on each of the values have been developed and extensively tested: the Schwartz Value Survey (Schwartz 1992) and the Portrait Values Questionnaire (PVQ) (Schwartz 2006).

Box 7.1 Schwartz's (1992) theory of basic human values

In 1992 Schwartz developed a theory of individual values that emerged from research investigating the possible universal content of human values in a cross-cultural context. Expanding and elaborating upon an earlier, smaller study, data for this research originated "from 40 samples in 20 countries" (Schwartz 1992, p. 3). The research addressed four basic issues: "What are the substantive contents of human values? Can we identify a comprehensive set of values? To what extent is the meaning of particular values equivalent for different groups of people? How are the relations among different values structured?" (Schwartz 1992, p. 59).

In above approach, "10 motivationally distinct value types [that are likely] recognized within and across cultures and used to form value priorities" (ibid.) were identified. From these ten overarching values, four broader and more abstract value-categories were identified: openness to change (self-direction, stimulation); conservation (security, conformity, tradition); self-transcendence (benevolence, universalism); and self-enhancement (hedonism, achievement, power). These values are portrayed within a circular model that illustrates the relationships and/or potential conflicts between each value (Schwartz 1992). Research evidence shows that "the meaning of the value types and of most of the single values that constitute them is reasonably equivalent across most groups" (Schwartz 1992, p. 59). Furthermore, "By identifying universal aspects of value content and structure [the research] has laid the foundations for investigating culture-specific aspects in the future" (Schwartz 1992, p. 60). In other words,

> against the background of common meanings and structure, it is now possible to compare the value priorities of cultures and groups and to detect genuine variation.... Unique, culture-specific understandings and applications of values will stand out against the universal patterns we have elucidated.
>
> (Ibid.)

Schwartz *et al.* (2012, p. 60) highlight the value of their theory as important for understanding "how the whole integrated system of value priorities relates to background, attitude, and behavior variables".

To describe the way in which grassroots innovations reflect and respond to the values of those who take part in them, we draw on a sociological perspective on values and organisations, given that grassroots innovations are organisations of one form or another. Chen *et al.* (2013, p. 857) identify organisations as one context "where values are collectively enacted or carried out.... Values may be discerned in any organization's goals, practices, and forms, including 'value-free' bureaucracies and collectivist organizations with participatory practices" (Chen *et al.* 2013, p. 856). Indeed, the authors suggest that far from being value-free, organisations in practice reflect, enact and propagate particular values. Accordingly, Martin and Upham (2016) argue that the mobilisation of values within the organisational structures of grassroots innovations processes can also be understood in terms of distinct processes of reflection, enactment and propagation (Chen *et al.* 2013).

In the above terms, the process of *reflection* on the outcomes, processes and structures of societal experiments reflects values, while *enactment* refers to the way in which societal experiments provide an opportunity for participants and activists to collectively enact both mainstream values and values that are less common. Furthermore, values can be enacted both through the objectives (ends) and the collective practices (means) of societal experiments. *Propagation* refers to the way in which values are propagated both within societal experiments and beyond their organisational boundaries. In both cases, institutional work – i.e. the efforts of "individual and collective actors aimed at creating, maintaining ... [or] disrupting institutions" (Lawrence *et al.* 2011, p. 52) – is undertaken to propagate novel practices, which themselves can propagate associated values. While Martin and Upham (2016) mostly focus on the mobilisation of Schwartz's ten basic values (Schwartz 1992; Schwartz *et al.* 2012), they also acknowledge that more complex and less widely held individual, organisational and cultural values may be at play in some regard.

A key premise of this study is that the rate and extent of the diffusion of grassroots innovation is in part determined (or influenced) by the predominant distribution of basic values across the general population. Hence, where a grassroots innovation such as online free reuse groups appeals mostly to people with very strong self-transcendence values (see Figure 7.1 and Table 7.1), the hypothesis in this case study is that the potential for diffusion of such grassroots innovation is likely to be limited by the relatively small number of people holding such values. In fact – and perhaps surprisingly – a strong mismatch in values is found *not* to be the case.

The case of Freegle

Online free reuse groups such as Freecycle and Freegle are based on the premise that "there is no such things as waste, it is just useful stuff in the wrong place" (Botsman and Rogers 2011, p. 124). These groups allow citizens, and to some extent organisations, the opportunity to give items they no longer need to other people in their local area. A digital platform provides the main medium of communication, and a wide range of items are offered to the users – usually at no cost (Groomes and Seyfang 2012). In these groups anyone from within a specified geographical area can join, thus limiting the distance people need to travel to collect the items (Botsman and Rogers 2011; Martin and Upham 2016). Each free reuse group is supported by local volunteers who facilitate the group activities (e.g. removing illegal or inappropriate posts, helping members with technical issues), and these volunteers generally promote reuse within their communities. Freegle was formed in 2009, when hundreds of volunteers concerned about the erosion of the grassroots ethos within Freecycle (Freegle 2014) broke away to form a new UK network (Jones 2009). Membership figures most likely overstate active participation, but in the UK there were 582 groups with 3.7 million members affiliated to Freecycle (Freecycle 2014) and 399 groups with 1.9 million members affiliated to Freegle in 2014 (Freegle 2014). Each network comprises a national umbrella organisation, local groups, organisation volunteers and group members.

There are alternatives to examining such activities from a psychological perspective. First, from an economic perspective, activities within free reuse groups can be seen as a form of generalised reciprocal exchange (Willer et al. 2012). From this perspective, individuals give items to other members of the group based on the implicit understanding that they, in the future, may draw upon the generosity of the group. Second, participants might also be viewed as engaging in a form of ethical consumption (Carrington et al. 2010), sustainable consumption (Young et al. 2010) or collaborative consumption (Botsman and Rogers 2011). Third, and more from a socio-logical perspective, the groups can be viewed as providing members with the opportunity to perform social practices that include reuse, gifting (Guillard and Bucchia 2012a) and anti-consumption (Black and Cherrier 2010). In this respect, the concepts of ethical citizenship (Schrader 2007) and ecological citizenship (Seyfang 2006) are also relevant (Martin and Upham 2016).

However, with each of the perspectives above there is a risk of ideal-ising the nature of free reuse groups and obscuring the varied motiva-tions of their members (Foden 2012; Martin and Upham 2016): judging by the material available on free reuse groups and posts on their web-sites, a wide range of factors motivate those participating in these groups. Such factors include: experiencing the pleasure associated with the act of making a gift (Nelson and Rademacher 2009); avoiding the inconveniences of other forms of waste disposal (Groomes and Seyfang 2012; Guillard and Bucchia 2012b); making a charitable gift to a person in need (Guillard and Bucchia 2012b; Nelson and Rademacher 2009); supporting the local community (Nelson et al. 2007); or acting on environmental concerns (Foden 2012). Motivations for requesting items include: saving money by acquiring items for free (Nelson et al. 2007); acting on environment concerns (Foden 2012); acquiring items to resell for profit or donate to charitable causes (Freecycle 2009); or acquiring items that the users could not otherwise afford (Foden 2012; Martin and Upham 2016). While Foden (2012) argues that participation in free reuse groups allows individuals to meet a need (i.e. either disposing of or acquiring an item) in a way that is consistent with their values, Nelson et al. (2007) argue that participants in these groups hold different consumption values to those engaged in mainstream consumer culture. Psychological perspectives explicitly address the topic of motiva-tion and the values and other constructs that are involved in behaviours such as engaging in Freegle or Freecycle. Hence the value of a psycho-logical perspective, alongside others.

Methods and results

Martin conducted an online survey to measure the values of Freegle group members. The *Portrait Values Questionnaire* (PVQ) (Schwartz 2006) was used in this survey, which consists of 21 specific questions that measure the emphasis placed by individual respondents on the values described in Table 7.1. The questions ask survey respondents to record how similar they are to a person portrayed in a short description (i.e. a portrait). Each portrait implicitly emphasises one of the ten basic values developed and applied by Schwartz and others over the last decades.[4] For instance, the portrait "He/she strongly believes that people should care for nature. Looking after the environment is important to him" emphasises the value of universalism. Responses were made on the following scale: 1 – "very much like me"; 2 – "like me"; 3 – "somewhat like me"; 4 – "a little like me"; 5 – "not like me"; 6 – not like me at all". The personal values of the survey respondents are thus inferred on the basis of similarity to the values of the portraits. Each PVQ question relates to one of ten basic values, and the responses were collated to create a raw score for each value for each survey respondent. These raw values scores were then centred as recommended by Schwartz (2006) – i.e. the mean of an individual's raw value responses is subtracted from each of their raw value scores, finally accounting for individual scoring tendencies.

In Martin's survey, questions on basic demographic data were also included, and the sampling approach was opportunistic or convenience-based; i.e. survey responses were sought from as many members of Freegle groups across the UK as possible – without stratification or the use of random sampling techniques. The final version of the survey gave 2,692 usable responses. The majority of respondents in the survey were female (67 per cent) and highly educated (i.e. held a university degree – 63 per cent), with a mean age of 53 years and a standard deviation of 13 years.

Having calculated the ten basic value scores for each survey respondent, a cluster analysis was undertaken to identify groups of survey participants with similar values, with a three cluster solution selected (Roussceuw 1987). The distribution of value scores for each of the clusters was then compared to the distribution of value scores of the UK population as measured by the European Social Survey.[5] Statistically significant differences between the mean value scores of survey respondents from the free reuse groups and the UK population overall were identified across *all* values ($p > 0.001$) using the independent samples t-test (Martin and Upham 2016). Overall, respondents from the reuse groups tended to emphasise the values of benevolence, universalism and self-direction, as shown by the negative mean value scores in Figure 7.2.

The cluster analysis of the respondents' values, identifying the three clusters as having distinct value profiles, is shown in Figure 7.3. Also, in this analysis the mean value scores for each cluster identified are shown as differing from the scores of the UK population in different ways.

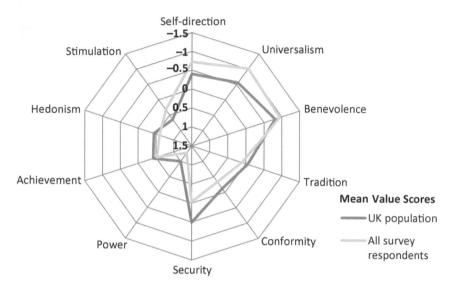

Figure 7.2 Mean basic values scores for the UK population and all survey respondents.

Note
Negative scores indicate that the value is emphasised by members of the sample; $n = 2,692$ for the Freegle survey and $n = 2,269$ for the UK population survey (European Social Survey 2012a); hierarchical cluster analysis in SPSS was used.

Cluster 1 includes 1,005 Freegle members (37 per cent of survey respondents) with a strong emphasis on self-transcendence (benevolence and universalism) and openness to change values (self-direction and stimulation). A comparison of mean value scores shows that cluster 1 members place a significantly stronger emphasis on both self-transcendence and openness to change values than members of the UK population overall.

Cluster 2 includes 616 Freegle users (23 per cent of survey respondents) with a strong emphasis on self-transcendence (benevolence and universalism) and conservation values (tradition, security and conformity). Again, comparison of mean value scores shows that cluster 2 members place a significantly stronger emphasis on both self-transcendence and conservation values than members of the UK population overall.

Cluster 3 includes 1,071 Freegle users (40 per cent of survey respondents). Compared to the other two clusters, this group emphasises self-transcendence values less (benevolence and universalism), but these respondents do place some emphasis on self-direction and security values. Although members of this cluster tend to emphasise self-transcendence values, they do so to a lesser degree than the UK population as a whole. Hence, and perhaps initially surprising, the composition of cluster 3 suggests that free reuse groups may appeal to citizens beyond those possessing

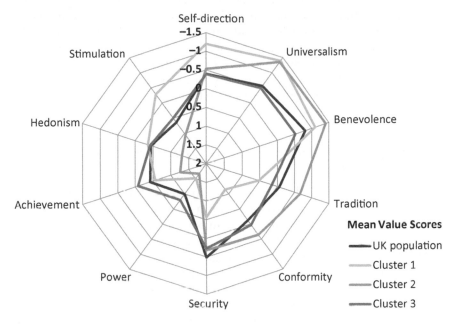

Figure 7.3 Mean basic values scores for the UK population and clusters 1, 2 and 3.

Note
Negative scores indicate that the value is emphasised by members of the sample; $n = 2,692$ in total for the Freegle survey and 2,269 for the UK population survey (European Social Survey 2012a); hierarchical cluster analysis in SPSS was used.

very strong pro-social (i.e. self-transcendence) values. This may imply that participation in pro-environmental grassroots innovation is not wholly dependent on pro-sociality. This, in turn, helps to explain how these groups have been able to grow beyond a small activist vanguard, reflected by the engagement (superficial or deep) of millions of citizens in free reuse groups in the UK (Martin and Upham 2016).

Discussion

Value enactment

The enactment of self-transcendence values is central to the dynamics of free reuse groups, in that the central action within the groups is giving an unwanted item freely to a stranger. Perhaps not surprisingly, across all three clusters the users of Freegle tend to express the self-transcendence values of universalism and benevolence, as shown in Table 7.1. We suggest that free reuse groups present affordances for participants to enact not only self-transcendence values, but also the values of openness to change and conservation (as emphasised in clusters 1 and 2 respectively). Furthermore, these affordances include opportunities to engage in action that is orientated towards social or sustainable development (self-transcendence); leads to new, novel or alternative experiences (openness to change); and/or resonates with personal and societal concepts of conservation in the broadest sense (conservation). In Table 7.1, we provide a brief overview of some of the means by which participants can enact basic values within free reuse groups.

Responding to and mobilising the values of free reuse group participants

The flexibility that enables participants to enact different values within these particular free reuse groups is not a chance occurrence. Free reuse activists *construct and maintain spaces* in which participants can meet their waste disposal and consumption needs in a way that is consistent with their values (Foden 2012). Indeed, Freegle activists seek to project a value-free or value- neutral image around the practice of free reuse. This is reflected in marketing on the website's homepage that makes minimal reference to environmental motivations.

This value-neutral image is constructed and maintained by reducing the explicit objective of the groups to reducing the amount of 'stuff' and waste sent to landfill. This objective arguably spans ideological and value-driven perspectives among the users, and it enables the groups to build a coalition of participants with a diverse range of values and motivations. However, while the core rules governing the groups are minimal, they do mandate that all items must be given away freely.

By establishing, supporting and actively using free reuse groups, free reuse activists mobilise the values of participants for particular and tangible ends. However, they also necessarily propagate some of the values of self-transcendence so central to the processes and structures of the groups. Furthermore, although free reuse groups do not have an explicit environmental or political agenda, some activists and participants do refer to concepts such as social and environmental justice (Nelson *et al.* 2007; Nelson and Rademacher 2009; Foden 2012). Hence, the redistribution of items from more affluent group members to economically and socially disadvantaged members, and the reduction in members' consumption levels/negative environmental impact, become desirable side effects of free reuse groups, towards which participant values are mobilised.

Implications for the diffusion of grassroots innovations

Rogers (1962, p. 5) defines diffusion as "the process by which an innovation is communicated through certain channels over time among the members of a social system". Here, a specific social psychological approach to values has been applied in order to examine the role of values in that diffusion, thus exploring the premise that the rate and extent of the diffusion of grassroots innovations through society is related – at least in part – to wider societal values. In the case of online free reuse groups such as Freegle and Freecycle, the distribution of values in the wider population do not appear to be a limiting factor for their development. On the contrary, the survey results show that many current reuse group participants (cluster 3: 40 per cent of the sample) have similar self-transcendence values to those of the general population. Moreover, it seems unlikely that the self-selecting nature of the survey sample would bias *against* self-transcendence; the converse is more likely (i.e. a more self-interested person may be less likely to complete a voluntary questionnaire than a more self-transcendent person).

However, this is not to argue that values are the only factors at play here. First, many other types of issues affect the scale-up of grassroots innovations. For example, as the groups are run by volunteers and reliant on generosity and trust between strangers, potential users may be deterred by the need to engage with a group instead of the anonymity of shopping, lack confidence in the groups or in the quality of the items offered (Vermeir and Verbeke 2008). Second, prevailing economic and political institutions limit the potential for pro-environmental behaviour generally (Blake 1999): the pressures to consume are strong. Third, scale-up may be limited by an incongruence in the *practices* of online free reuse groups, relative to the prevailing, habitual and routine practices of consumption and waste disposal (Hargreaves *et al.* 2013b), which are themselves supported by institutionalised procedures that together exemplify the structural challenges faced by grassroots innovations (Seyfang and Smith 2007). The interplay of practices and values itself merits further attention (Piscicelli *et al.* 2015).

Online free reuse groups represent a form of sociotechnical change that to some extent conflicts with prevailing regimes of consumption/production and the associated social practices. Even if there is a (perhaps surprising) degree of value overlap with wider society, free reuse groups – and possibly many other grassroots innovations – enact, seek to propagate and express both certain marginal and marginalised values. As suggested in the multi-level perspective (MLP) framework, politics and power are bound up in these processes, with active resistance by the prevailing regimes (Geels 2014), which tend to enact rather different values than those mobilised through specific niche-level grassroots innovations. In this context, for example, such regime-level values enact high levels of consumption required by systems of provision dependent on economic growth and material consumption. Extending the use period (life) of products through sharing is contradictory to this.

Further considerations

The analytical approach suggested above embodies both concepts and causal processes intended to help analyse, characterise and explain the mobilisation of values within online free reuse communities. This approach raises a number of issues and thoughts on the further development and application of value-based approaches in relation to grassroots innovations. We highlight these below (we also follow up on broader research directions beyond grassroots innovations in Chapter 9).

Focusing on basic values has proved useful as an analytical device, obliging us to be explicit about which values in an innovation process are being mobilised and by whom. However, grassroots innovation research suggests that other, non-basic values as well as more specific attitudes (e.g. deep green values and anti-consumption attitudes respectively) can also play an important role in driving societal experiments (e.g. Seyfang and Smith 2007; Seyfang *et al.* 2014). Furthermore, where participants use free reuse groups to engage in the practices of ethical consumption (Foden 2012), it may be that anti-capitalist values (Shaw *et al.* 2005) and ecological citizenship values (Seyfang 2006) also play important motivational roles. In other words, we would not claim to have tapped the full range of values that might be operating in this context, let alone in others relevant to sociotechnical sustainability transitions.

Moreover, from a practice-based perspective, psychological constructs such as attitudes are viewed as conditional on practices, rather than vice versa (Shove 2010). Sometimes values might follow practices, and so perhaps existing practices may be even more important than values for the diffusion of grassroots innovations. From a practice theory perspective, attitudes are seen as physically, cognitively and emotionally integrated into ways of living, and practices are also seen as integrated in and connected to the multiplicities of arrangements that make up lifestyles (Warde 2005). From this perspective, then, those involved in promoting the diffusion of grassroots innovations need to consider: how and what values are projected via those innovations; what types of practice are prevalent in the related contexts (Hargreaves *et al.* 2013b); and what is required by these practices (ibid.).

The research approach presented above also holds the potential to integrate other aspects of sociotechnical transitions theory more explicitly. For example, research might consider which values are enacted and propagated by the prevailing sociotechnical systems that serve specific societal needs (i.e. the regime level of the MLP; Geels 2005), thus exploring how such dynamics may either limit or open up opportunities for grassroots and other forms of innovation. In particular, there is scope here for considering how values relate to dynamics in the broader conceptions of sociotechnical change. For example, De Haan and Rotmans (2011) conceive some of the particular conditions required for such sociotechnical transitions as:

(a) cultural and structural tensions; (b) a degree of internal inconsistency (stress); and (c) pressures from inside or outside the regime. These authors also speak of particular, sequential patterns or processes that transitions undergo, such as empowerment, reconstellation and adaptation: as socio-technical constellations build in strength, they become materially and cognitively installed and the regime form shifts to accommodate the innovation. Indeed, as Leiserowitz *et al.* (2006) identify, a fundamental change is needed in how societies prioritise values that are consistent with the transition to sustainability. However, it is also possible that the scope of environmental protection can be increasingly broadened and understood as an expression of social altruistic values, representing more of a conceptual or cognitive change than a value change.

Conclusion

There is growing interest in civil society as a site from which 'grassroots social innovations' may emerge, and such grassroots innovations may contribute significantly to the transition to more sustainable production and consumption systems. Drawing upon Martin and Upham (2016), this chapter offers a conceptual model of the role that values may play in grassroots innovations as such initiatives emerge from the niche, possibly entering the regime. In this study, a large-scale survey of UK online free reuse groups applied psychological value scales to assess the values of the free reuse group users. The study found that while the values of self-transcendence (benevolence and universalism) were emphasised by a majority of the participants to a significantly greater degree than in the UK population as a whole, a large minority (40 per cent) actually emphasised self-transcendence values to a lesser degree than the UK population overall. Moreover, those participants who did emphasise self-transcendent values were not mono-dimensional in their value sets, but also held other values of significance to the wider population. While this is to be expected, there is surprisingly little work on the role of values in relation to concepts of sociotechnical transitions or of grassroots innovations as a feature of these transitions. Indeed, this study could raise more questions than it provides answers, hence opening up possibilities for several novel research directions specific to grassroots innovations. We address these in Chapter 9, on sociotechnical transitions dynamics more generally.

Notes

1 Martin, C. J. and Upham, P. (2016). "Grassroots social innovation and the mobilisation of values in collaborative consumption: A conceptual model". *Journal of Cleaner Production* 134:204–13. Data collection and statistical analysis was by Chris Martin, research design and interpretation by Martin and Upham.

2 Actually we could not find any such papers in Mokter's (2018) review of 87 sifted grassroots innovation articles, but perhaps they are out there somewhere.
3 In terms of an empirical test of this, Rudnev *et al.* (2018) examine the strength of correlation between the value groups in Schwartz's scale, using data from 104 countries. They find the theoretically expected negative relations between openness to change and conservation values; and between self-transcendence and self-enhancement values. That is, across most countries, values tend to be organised predominantly in line with the social versus person focus opposition (others versus oneself). The growth versus self-protection opposition is pronounced only in the more economically developed countries (Rudnev *et al.* 2018).
4 At the time of writing, Schwartz's most recent statement of his theory of basic values per se is Schwartz (2017). Note that, arguably, one should not conflate individual, personality-level values and society-level, cultural values (Schwartz 2011).
5 The European Social Survey (ESS), conducted in 2012, included the PVQ (European Social Survey 2012b) and was completed by a representative sample of the UK population (European Social Survey 2012a) – 2,269 people over the age of 15.

Bibliography

Bakker, C., Wang, F., Huisman, J. and den Hollander, M. 2014. "Products that go round: Exploring product life extension through design". *Journal of Cleaner Production* 69:10–16. doi:10.1016/j.jclepro.2014.01.028.

Battilana, J., Leca, B. and Boxenbaum, E. 2009. "How actors change institutions: Towards a theory of institutional entrepreneurship". *The Academy of Management Annals* 3:65–107.

Black, I. R. and Cherrier, H. 2010. "Anti-consumption as part of living a sustainable lifestyle: Daily practices, contextual motivations and subjective values". *Journal of Consumer Behaviour* 9:437–53.

Blake, J. 1999. "Overcoming the 'Value–Action Gap' in environmental policy: Tensions between national policy and local experience". *Local Environment* 4:257–78.

Botsman, R. and Rogers, R. 2011. *What's Mine is Yours: How Collaborative Consumption is Changing the Way We Live.* London: Collins.

Boyer, R. 2014. "Sociotechnical transitions and urban planning: A case study of eco-cohousing in Tompkins County, New York". *Journal of Planning Education and Research* 34:451–64.

Carrington, M., Neville, B. and Whitwell, G. 2010. "Why ethical consumers don't walk their talk: Towards a framework for understanding the gap between the ethical purchase intentions and actual buying behaviour of ethically minded consumers". *Journal of Business Ethics* 97:139–58.

Chen, K. K., Lune, H. and Queen, E. L. 2013. "How values shape and are shaped by nonprofit and voluntary organizations: The current state of the field". *Nonprofit and Voluntary Sector Quarterly* 42:856–85.

De Haan, J. and Rotmans, J. 2011. "Patterns in transitions: Understanding complex chains of change". *Technological Forecasting and Social Change* 78:90–102.

Dendler, L. 2014. "Sustainability meta labelling: An effective measure to facilitate more sustainable consumption and production?". *Journal of Cleaner Production* 63:74–83.

Dietz, T., Fitzgerald, A. and Shwom, R. 2005. "Environmental values". *Annual Review of Environment and Resources* 30:335–72.

European Social Survey. 2012a. "Sampling for the European Social Survey Round VI: Principles and requirements". Available at: www.europeansocialsurvey.org/docs/round6/methods/ESS6_sampling_guidelines.pdf [accessed 19 November 2014].

European Social Survey. 2012b. *United Kingdom – documents and data files* [online]. Available at: www.europeansocialsurvey.org/data/country.html?c=united_kingdom [accessed 19 November 2014].

Foden, M. 2012. "Everyday consumption practices as a site for activism? Exploring the motivations of grassroots reuse groups". *People, Place & Policy Online* 6:148–63.

Freecycle. 2009. *When Members Solicit Sales in Response to a WANTED Post* [online]. Available at: http://wiki.freecycle.org/When_Members_Solicit_Sales_in_Response_to_a_WANTED_Post [accessed 13 March 2015].

Freecycle. 2014. *The Freecycle Network* [online]. Available at: www.freecycle.org/ [accessed 11 November 2014].

Freegle. 2014. *About Freegle* [online]. Available at: www.ilovefreegle.org/about/ [accessed 19 November 2014].

Geels, F. W. 2005. "The dynamics of transitions in sociotechnical systems: A multi-level analysis of the transition pathway from horse-drawn carriages to automobiles (1860–1930)". *Technology Analysis & Strategic Management* 17:445–76.

Geels, F. W. 2014. "Regime resistance against low-carbon transitions: Introducing politics and power into the multi-level perspective". *Theory, Culture & Society* 31:21–40.

Giddens, A. 1984. *The Constitution of Society: Outline of the Theory of Structuration*. Cambridge: Polity Press.

Groomes, L. and Seyfang, G. 2012. "Secondhand spaces and sustainable consumption: Examining Freecycle's environmental impacts and user motivations". Science, Society and Sustainability (3S) Research Group, University of East Anglia, UK. Available at: https://uea3s.files.wordpress.com/2014/12/3s-wp-2012-05-groomes-and-seyfang.pdf [accessed 28 November 2018].

Guillard, V. and Bucchia, C. D. 2012a. "How about giving my things away over the internet? When internet makes it easier to give things away". In: Z. Gürhan-Canli, C. Otnes and R. J. Zhu (eds.), *Advances in Consumer Research*. Duluth, MN: Association for Consumer Research.

Guillard, V. and Bucchia, C. D. 2012b. "When online recycling enables givers to escape the tensions of the gift economy". In: R. W. Belk, S. Askegaard and L. Scott, (eds.), *Research in Consumer Behavior*. Bingley, UK: Emerald Group Publishing.

Hargreaves, T., Hielscher, S., Seyfang, G. and Smith, A. 2013a. "Grassroots innovations in community energy: The role of intermediaries in niche development". *Global Environmental Change* 23:868–80.

Hargreaves, T., Longhurst, N. and Seyfang, G. 2013b. "Up, down, round and round: Connecting regimes and practices in innovation for sustainability". *Environment and Planning A* 45:402–20.

Heiskanen, E., Jalas, M., Rinkinen, J. and Tainio, P. 2015. "The local community as a 'low-carbon lab': Promises and perils". *Environmental Innovation and Societal Transitions* 14:149–64.

Hitlin, S. and Piliavin, J. A. 2004. "Values: Reviving a dormant concept". *Annual Review of Sociology* 30:359–93.

Hossain, M. 2018. "Grassroots innovation: The state of the art and future perspectives". *Technology in Society* 55:63–9. doi:10.1016/j.techsoc.2018.06.008.

Jackson, T. 2005. "Motivating sustainable consumption. A review of evidence on consumer behavior and behavioral change". Centre for Environmental Strategy, University of Surrey, UK. www.sustainablelifestyles.ac.uk/sites/default/files/motivating_sc_final.pdf.

Jones, S. 2009. "Accusations of very tight control split UK recycling network from US parent". *Guardian* [online], 12 October 2009. Available at: www.theguardian.com/environment/2009/oct/12/freecycle-freegle-recycling-networks-groups [accessed 19 November 2014].

Kirwan, J., Ilbery, B., Maye, D. and Carey, J. 2013. "Grassroots social innovations and food localisation: An investigation of the Local Food programme in England". *Global Environmental Change* 23:830–7.

Lawrence, T., Suddaby, R. and Leca, B. 2011. "Institutional work: Refocusing institutional studies of organization". *Journal of Management Inquiry* 20:52–8.

Leiserowitz, A. A., Kates, R. W. and Parris, T. M. 2006. "Sustainability values, attitudes, and behaviors: A review of multinational and global trends". *Annual Review of Environment and Resources* 31:413–44.

Lind, H. B., Nordfjærn, T., Jørgensen, S. H. and Rundmo, T. 2015. "The value-belief-norm theory, personal norms and sustainable travel mode choice in urban areas". *Journal of Environmental Psychology* 44:119–25. doi:10.1016/j.jenvp.2015.06.001.

Markard, J., Raven, R. and Truffer, B. 2012. "Sustainability transitions: An emerging field of research and its prospects". *Research Policy* 41:955–67.

Martin, C. J. and Upham, P. 2016. "Grassroots social innovation and the mobilisation of values in collaborative consumption: A conceptual model". *Journal of Cleaner Production* 134:204–13. doi:10.1016/j.jclepro.2015.04.062.

Mokter, H. 2018. "Grassroots innovation: The state of the art and future perspectives". *Technology in Society* 55:63–9. doi:10.1016/j.techsoc.2018.06.008.

Nelson, M. R. and Rademacher, M. A. 2009. "From trash to treasure: Freecycle.org as a case of generalized reciprocity". *Advances in Consumer Research* 36:905–6.

Nelson, M. R., Rademacher, M. A. and Paek, H-J. 2007. "Downshifting consumer = upshifting citizen? An examination of a local Freecycle community". *The Annals of the American Academy of Political and Social Science* 611:141–56.

Piscicelli, L., Cooper, T. and Fisher, T. 2015. "The role of values in collaborative consumption: Insights from a product-service system for lending and borrowing in the UK". *Journal of Cleaner Production* 97:21–9.

Putnam, R. D. 2000. *Bowling Alone*. New York: Simon and Schuster.

Rogers, E. M. 1962. *Diffusion of Innovations*, New York: Free Press.

Rousseeuw, P. J. 1987. "Silhouettes: A graphical aid to the interpretation and validation of cluster analysis". *Journal of Computational and Applied Mathematics* 20:53–65. doi:10.1016/0377-0427(87)90125-7.

Rudnev, M., Magun, V. and Schwartz, S. 2018. "Relations among higher order values around the world". *Journal of Cross-Cultural Psychology* 49(8):1165–82.

Schrader, U. 2007. "The moral responsibility of consumers as citizens". *International Journal of Innovation and Sustainable Development* 2:79–96.

Schwartz, S. H. 1977. "Normative influences on altruism". *Advances in Experimental Social Psychology* 10:221–79. doi:10.1016/S0065-2601(08)60358-5.

Schwartz, S. H. 1992. "Universals in the content and structure of values: Theoretical advances and empirical tests in 20 countries". *Advances in Experimental Social Psychology* 25:1–65.

Schwartz, S. H. 2006. "Les valeurs de base de la personne: Théorie, mesures et applications". *Revue Française de Sociologie* 47:929–68.

Schwartz, S. H. 2011. "Values: Individual and cultural". In: F. J. R. van de Vijver, A. Chasiotis and S. M. Breugelmans (eds.), *Fundamental Questions in Cross-Cultural Psychology*. Cambridge: Cambridge University Press, pp. 463–93.

Schwartz, S. H. 2017. "The refined theory of basic values". In: S. Roccas and L. Sagiv (eds.), *Values and Behavior*. Cham, Switzerland: Springer.

Schwartz, S. H. and Bilsky, W. 1987. "Toward a universal psychological structure of human values". *Journal of Personality and Social Psychology* 53:550–62.

Schwartz, S. H., Cieciuch, J., Vecchione, M., Davidov, E., Fischer, R., Beierlein, C., Ramos, A., Verkasalo, M., Lönnqvist, J-E., Demirutku, K., Dirilen-Gumus, O. and Konty, M. 2012. "Refining the theory of basic individual values". *Journal of Personality and Social Psychology* 103:663–88.

Seyfang, G. 2006. "Ecological citizenship and sustainable consumption: Examining local organic food networks". *Journal of Rural Studies* 22:383–95.

Seyfang, G. and Longhurst, N. 2013. "Desperately seeking niches: Grassroots innovations and niche development in the community currency field". *Global Environmental Change* 23:881–91.

Seyfang, G. and Smith, A. 2007. "Grassroots innovations for sustainable development: Towards a new research and policy agenda". *Environmental Politics* 16:584–603.

Seyfang, G., Hielscher, S., Hargreaves, T., Martiskainen, M. and Smith, A. 2014. "A grassroots sustainable energy niche? Reflections on community energy in the UK". *Environmental Innovation and Societal Transitions* 13:21–44.

Shaw, D., Grehan, E., Shiu, E., Hassan, L. and Thomson, J. 2005. "An exploration of values in ethical consumer decision making". *Journal of Consumer Behaviour* 4:185–200.

Shove, E. 2010. "Beyond the ABC: Climate change policy and theories of social change". *Environment and Planning A* 42:1273–85.

Smith, A. and Raven, R. 2012. "What is protective space? Reconsidering niches in transitions to sustainability". *Research Policy* 41:1025–36.

Smith, A., Voß, J-P. and Grin, J. 2010. "Innovation studies and sustainability transitions: The allure of the multi-level perspective and its challenges". *Research Policy* 39:435–48.

Smith, A., Fressoli, M. and Thomas, H. 2014. "Grassroots innovation movements: Challenges and contributions". *Journal of Cleaner Production* 63:114–24.

Star, S. L. and Griesemer, J. R. 1989. "Institutional ecology, 'translations' and boundary objects: Amateurs and professionals in Berkeley's Museum of Vertebrate Zoology, 1907–39". *Social Studies of Science* 19:387–420.

Stern, P. C. and Dietz, T. 1994. "The value basis of environmental concern". *Journal of Social Issues* 50:65–84.

Stern, P. C., Dietz, T. and Guagnano, G. A. 1995. "The new ecological paradigm in social-psychological context". *Environment and Behavior* 27:723–43.

Stern, P. C., Dietz, T., Abel, T., Guagnano, G. A. and Kalof, L. 1999. "A value-belief-norm theory of support for social movements: The case of environmentalism". *Human Ecology Review* 6:81–98.

Vergragt, P., Akenji, L. and Dewick, P. 2014. "Sustainable production, consumption, and livelihoods: Global and regional research perspectives". *Journal of Cleaner Production* 63:1–12.

Vermeir, I. and Verbeke, W. 2008. "Sustainable food consumption among young adults in Belgium: Theory of planned behaviour and the role of confidence and values". *Ecological Economics* 64:542–53.

Warde, A. 2005. "Consumption and theories of practice". *Journal of Consumer Culture* 5:131–53.

Willer, R., Flynn, F. J. and Zak, S. 2012. "Structure, identity, and solidarity: A comparative field study of generalized and direct exchange". *Administrative Science Quarterly* 57:119–55.

Young, W., Hwang, K., McDonald, S. and Oates, C. J. 2010. "Sustainable consumption: Green consumer behaviour when purchasing products". *Sustainable Development* 18:20–31.

8 Sociotechnical transitions governance and public engagement

Introduction

In the previous chapters, we have addressed ways in which various psychological perspectives can be more or less closely connected to sociotechnical transitions theory, but we have not considered issues of public engagement in – and consultation on – such transition policies and their development. In general, both in research and among practitioners of environmental management and planning, including energy infrastructure planning, some form of public engagement and consultation are perceived as desirable (Raven *et al.* 2012). Yet among the multiple stakeholders in such processes, what forms these processes should take, and what their formal role should be in decision-making and policy processes has been greatly contested. All of this is strongly connected to issues of power, which have been given increasingly (but arguably still insufficient) explicit attention in the transitions governance literature (Avelino and Rotmans 2009; Avelino and Wittmayer 2016).

This chapter draws on some of these issues through a study, by Upham *et al.* (2015), of public perceptions of low carbon transport innovation policy options in Finland. It does so leaning on the results of a large-scale survey on the Finnish land-based passenger transport sector, as this might develop for a lower carbon future. More broadly, this study focuses on and discusses the engagement of publics as key stakeholders in sociotechnical transition processes. In this regard, the notion and role of transitions management (TM) is centrally discussed.

Overall, the study highlights the differentiated nature of public support for innovation policy for low carbon transport, and it suggests that accounting for these differences is a precondition for securing the broader social legitimacy of TM processes. Drawing upon the survey-based evidence from within this one exemplar country, we show how transport practices and attitudes to transport innovation policy vary widely with both demography and geography, and discuss the way in which differences in public opinion might hinder or complicate transition-related policy-making. We argue that such knowledge matters for the effectiveness – or at least legitimacy – of policy design.

We also suggest that particular forms of public opinion surveying can provide a relatively inexpensive way of obtaining information about the (potential) sub-national and regional differences in public opinion in relation to specific initiatives. Such knowledge may be deepened through iteration: i.e. using the responses from one survey to refine questions in follow-up surveys; through the involvement of specific, potentially impacted population groups in the design of questions; and, more generally, through the use of the various methods of opinion elicitation commonly used in the social sciences and in political research. This in turn raises many normative issues, which we touch on, including the relative weight that should or might be given to the views of technical experts, lay publics, elected political representatives, state agencies and commerce in policy and TM processes.

Transitions management

Overall, the ambition of TM implies moving beyond just stakeholder *consultation* in policy-making, towards engaging stakeholders actively in these processes, giving weight to their views and working towards more sustainable prescriptions for policy and practice that have wide societal support. In this way, TM is intended as a socially participative approach to the governance of sociotechnical transitions towards environmental sustainability (Kemp *et al.* 2007; Loorbach and Rotmans 2010). Conceptually, TM is conceived of as a form of multi-level governance in which state and non-state actors are brought together to co-produce policies with the aim of co-ordinating science, innovation and sectoral policy (Kemp *et al.* 2007). While TM offers analytic concepts and descriptive characterisation of sociotechnical change processes, the approach is also prescriptive, offering designs for sustainability governance (Loorbach 2009). It departs from (and extends) the innovation studies literature by its explicitly normative stance, and it adopts sustainability as an explicit objective. Practical and theoretical work with TM has continued most strongly in the Netherlands (e.g. Frantzeskaki and de Haan 2009; Loorbach and Rotmans 2010; de Haan and Rotmans 2011; Frantzeskaki *et al.* 2012), but it has also drawn on empirics relating to transitions experiments elsewhere (e.g. Nevens *et al.* 2013; Nevens and Roorda 2014). Conceptually, the approach is rooted in innovation studies (Geels and Schot 2007); a modest number of academic papers that are centrally about TM are published annually, with a mean of 15 papers per year published between 2015 and 2018.[1]

Public participation in sociotechnical innovation

For several decades, science and technology studies (STS) theorists have argued for more authentic and deeper public engagement in technology innovation research processes, arguing that this would serve as a means of enhancing the legitimacy of innovation processes (e.g. Wynne 1973; Sclove 1995). Indeed, 'public participation' and 'local engagement' have long been on the agenda of many specific practical and academic domains – including in the transportation planning domain that is of relevance here (e.g. Wellman 1978).

Such 'participation' has long been discussed in multiple contexts and in all senses of 'the public', from individuals or local publics, through to civil society in a variety of settings, from technology design through to technology use (Nahuis and van Lente 2008). Yet the question of *how* to meaningfully engage publics in processes of both scientific and technological innovation remains (Macnaghten and Chilvers 2014). Moreover, all too often this question is addressed in very general terms that do not address the importance of local heterogeneities or acknowledge the power-related issues involved. From the perspective of policy legitimacy theory, public participation in policy-making, and the subsequent demonstrable transparency of those processes, should support and promote stakeholder perceptions of procedural legitimacy (Suchman 1995). In this context, legitimacy can be broadly defined as: "a generalized perception or assumption that the actions of an entity are desirable, proper, or appropriate within some socially constructed system of norms, values, beliefs, and definitions" (Suchman 1995, p. 574). Procedural legitimacy of policy thus relates to the *process* by which this policy is made, while stakeholder perceptions of the more general legitimacy of the State rests on their overall beliefs in its rightfulness and moral authority (Barker 1990; Upham and Dendler 2014). Here, we address the question of how to achieve and promote procedural policy legitimacy for innovation policy via public engagement, and we do this from the perspective of TM. Our premise is that there are different degrees of legitimacy, and that these differing degrees may be achieved among different groups within society.

There is another issue that is relevant here: geography. In its search for generally applicable theories and concepts, sociotechnical transitions theory has been critiqued for inadequately addressing the spatial/geographical aspects and specificities of transitions dynamics (Coenen *et al.* 2012). As we show below, geography, or more precisely the very specific local contexts in which transitions processes take place, can be important for the character of public opinions towards those changes in local areas – and hence also for the results of efforts to create policy legitimacy regionally. Efforts at introducing geographical dimensions in sociotechnical studies include those by Coenen *et al.* (2012) and Raven *et al.* (2012). These authors have emphasised the importance of relational, network-based aspects of transition: actor networks that are made possible and reinforced by co-location, that is, locations in in some way connected to one another, typically through physical proximity. While there is some work on introducing spatiality into transitions thinking, we see little or no work in the transitions literature on the role of geographically-based differences in public opinion, the implications of such differences for 'managing' these transitions and implications for national (or regional) policy.

In this study, our focus emphasises the need to account for geographical and other differences among 'the public' or publics when seeking to engender policy legitimacy in transitions processes. With the exemplar sector of land-based passenger transport, the chapter discusses the nature of differences in public opinions of state policy options for sociotechnical innovations, and it specifically addresses the way in which such differences are geographically influenced. Leaning on these research insights, the implications of differences in opinions among publics for governing the transition of the transport sector are discussed – particularly regarding public engagement and its legitimating role for these processes, and regarding which types of sociotechnical innovation may be favoured and accepted. In this respect, we argue that heterogeneity in public opinion has implications for achieving policy legitimacy and that this is obscured by a focus on broader majority opinion alone. Accordingly, we suggest that the iterative and participative use of opinion surveys can play *a* role in engaging the public in the social participation that TM envisages, but only provided that these are undertaken in a manner that is mindful of the power relations involved and with transparency regarding the degree of influence that the participating publics may expect.

Theory

Transition management and public engagement in innovation policy

TM is a response to the view that neither central planning by governments nor market forces are sufficient to bring about the types of change that complex, persistent and interconnected social, economic and environmental problems require (Loorbach 2009). As an approach to governance, TM proposes and applies "an instrumental, practice-oriented model" in order to influence ongoing sustainability transitions through reflexive and evolutionary governance (Markard *et al.* 2012). The perspective is systems-based: it assumes that social, technological, economic and other phenomena and actors are connected more or less directly, and therefore that interventions or pressures at one point in the system may induce or influence change elsewhere in the system.

The need for reflexivity that TM emphasises is in part analytical, but also normative in its underlying (if tacit) democratic principles. Moreover, in TM the networks that social, economic and governmental actors form and foster are not treated uncritically, but seen as potentially (Hendriks 2008, 2009) involving limited transparency and accountability (Loorbach and Rotmans 2010). As an ideal, TM is an attempt to increase the broader social legitimacy and effectiveness of new forms of governance by: (a) offering a structural perspective on system change that itself builds on a multi-level perspective (MLP) of interconnected social and technological change (Rip and Kemp 1998); and (b) proposing and testing new fora and methods of governance, particularly bringing together different actors and/or actor perspectives on transitions (Upham *et al.* 2015).

Some key concepts from TM have been outlined above, but there are further issues that should be considered before setting out a basic proposal for public engagement in TM. For this purpose, Upham *et al.* (2015) draw selectively on a categorisation of engagement issues by Delgado *et al.* (2011). First, *why* pursue public engagement at all? In a very instrumental sense, early public engagement seeks to minimise or, indeed, eliminate opposition towards the projects or processes at hand by attending to and addressing questions and queries from publics and/or other stakeholders at the earliest stage possible. In some instances, dependent on and reflecting the project type and scope, public engagement and participation initiatives are open, explorative and creative, and some may sport an ideal of collaborative project design and a relative balance of power between the parties involved. Arguably, however, such cases are rare. Public engagement in its most limited form may consist of public consultations, and these are more common. While many organised public engagement processes in scientific and technological development are arguably not intended as mere consultations, but also as processes of two-way communication, inspiration and dialogue, sometimes these activities nonetheless resemble one-way information provision, i.e. consultations. Despite the limitations of such consultations (intentional or not), effective and well-planned one-way dissemination of information still does and will continue to play a role in the process of generating socially shared visions of the future in TM and processes of policy-making.

Second, *who* should participate, i.e. which 'publics', which segments of the population, and how and by whom are these participants recruited? These are also pivotal questions in any public engagement and participation activities. If broad public representation is a given objective in engagement endeavours, then stratified polling (the systematic selection of respondents that provides a demographically representative sample of the selected population) is ideally required (Rowe and Frewer 2000). Sometimes, narrower demographic or geographic population segments may be relevant for the participation processes and, in such cases, more exploratory work is perhaps merited. Hence, a broad or narrow scoping of surveys can, for example, be used and qualitative work (interviews, focus groups, ethnography, etc.) may also help to reveal themes or characteristics of interest that can then be followed up with additional survey work.

Third, *how* the terms of engagement are decided upon may vary. Initially, deciding upon relevant topics and questions of interest can be determined in partnership with affected populations and/or other relevant stakeholders. Such discussions will likely increase the legitimacy of the results and outcomes due to the enhanced transparency of the process (Rowe and Frewer 2000).

Fourth, *when* the engagement should take place (at what point in the development of a given process it takes place) will depend on the purpose, context and constraints of the situation. An affected population may prefer early and ongoing involvement in the decision processes, and indeed maximum influence over those processes, but this may or may not be politically and institutionally possible (or perhaps desirable, depending on one's view – and on the technical/legal requirements of a particular project). In principle, early engagement gives more time to shape decision-making. Ultimately, however, 'engagement' in decision-making processes at any stage is only true engagement if there is full transparency as to the actual level of influence on the outcomes of (perhaps certain parts of) those decisions that is practically possible – and/or desirable.

Figure 8.1 brings these considerations together with the ideas of TM. First, increased public engagement may or may not, on the one hand, facilitate a particular transition, not least because increased debate may also complicate argument closure. On the other hand, insufficient debate may lead to additional dissent and objections among members of the public that it may have been easier to deal with at an earlier stage in the innovation process. Second, public opinion on the transition processes influences transition drivers and dynamics at all stages, positively and negatively, and this influence will persist regardless of any ambitions for TM – inclusive of participation processes. Moreover, members of the heterogeneous public, or publics, will sometimes emphasise their different roles – for example, as citizens, consumers or civil society actors with a more consciously political role.[2]

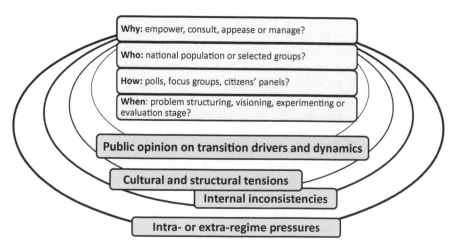

Figure 8.1 Public engagement and transition management processes.

Having outlined some differing rationales for public engagement in TM and some basic methods for how this might take place, we now draw on a case study by Upham *et al.* (2015) for illustration. This study shows how public opinion on low carbon innovation policy options for the land-based passenger transport sector in Finland is demographically and geographically differentiated. The work is experimental, rather than part of a live policy process, but nonetheless the detailed empirics illustrate the types of opinion difference that may influence a future TM process.

Case study

The research design is exploratory and inductive, the aim being to explore the governance implications of diverse public opinions for public engagement relating to specific transition processes. The selected samples of public opinion are from three different work-to-travel areas in Finland. The specific objectives of the study were: (1) to reveal public opinion on transport innovation and innovation policy options; (2) to identify the nature of any differences within the selected populations; and (3) to provide an empirical basis for further discussion of public engagement in TM processes.

The survey was part of a wider project on low carbon, system level transitions for transport (Temmes *et al.* 2014). Public opinion was elicited with an online survey instrument designed to take 20–30 minutes to complete, and the full sample consisted of 1,000 people split equally across the contrasting travel to work areas (TTWAs). The survey was administered by a market research firm. In each TTWA, demographic representation was sought in terms of gender, age and social class. The TTWAs selected were the Helsinki, Tampere and Oulu regions of Finland. These represent a capital city with an extensive public transport system including metro, trams and buses (Helsinki); a regional city with a bus system (Tampere); and a smaller, more peripherally- and rurally-located city with a bus system (Oulu). In the study, the selected TTWAs were assumed to have differing patterns of transport use, mirroring the different public transport infrastructure availability in those places and possibly different perceptions of innovation priorities that reflected different economic interests and outlooks among the respondents. In particular, it was hypothesised that views in Oulu would differ from those in Helsinki and Tampere and, as we show, this was indeed found to be the case.

Question design represented a range of technological, behavioural and legislative change and transition options, reflecting previous transitions work on transport (Geels 2012) and on 'sustainable' transport policy options (Banister 2008). In particular, both 'soft' and 'hard' transport policy and innovation options were included: those that involved social and institutional innovation (soft), as well as those in which innovation is primarily technological (hard). Other relevant opinion surveys were also taken into account (e.g. EC 2011), as well as relevant reports on future transport and mobility options (McKinsey & Company 2009; VTPI 2010; VTT 2012; European Climate Foundation 2013; PE International 2013). The questions also reflected the way in which "[t]ransport and travel choices are rooted in the structure of activities undertaken by individuals and families", and "attitudes to transport must also be rooted in deeper values and aspirations of how people want to lead their lives" (Goodwin and Lyons 2010, p. 16). Hence questions on respondents' everyday transport habits and practice and on environmental attitudes/values were also included in the questionnaire.

A key premise of the survey design was that different degrees of dependence on differing transport modes may affect attitudes to transport innovation policy options and, in a similar vein, that differences between the selected TTWAs, such as prevalent types of economic activities in these locations, may also have an effect. Helsinki, on the southern coast of Finland, is a cosmopolitan capital city with an extensive bus, tram and metro network and a climate that, while hardly mild in winter, is milder than further north. Tampere is a regional city 90 minutes north of Helsinki by train, and it shares many of the same characteristics as Helsinki, albeit on a smaller scale. In contrast, Oulu is climatically sub-arctic. This city has shorter internal transport distances, and it has a smaller mixed economy that combines high tech start-ups with port logistics and materials processing (notably wood and steel).

The survey results regarding the different ways in which the Finnish state might support low carbon innovation policy for sustainability transition in personal transport show significant opinion differences that are linked to both demography and geography. Table 8.1 highlights some of the implications of the differences for public policy legitimacy, and it does so in Suchman's (1995) broad sense of the word 'legitimacy'. Table 8.1 distinguishes between aggregate opinion, presented first, and opinion differentiated by geography, gender, age and income.

Earlier in the chapter, we argued that there are multiple choices to be made regarding the motives and modes of public engagement in transition processes. We also argued that public opinion is likely to impinge on transition processes in any case. In Table 8.1, we can see examples of how this influence might become manifest in terms of citizen support for public policy with the example of the transport sector. First, at an aggregate level the sample population (1,000 people) can be characterised as somewhat conservative, valuing incremental technological innovations such as more fuel-efficient conventional vehicles, biofuels and hybrids, but valuing public transport related investments too.[3] Most respondents accept that changes are required to mitigate climate change impacts, but a significant minority do not. Climate scepticism, defined in the questionnaire as disagreeing that "the world's climate is changing due to human activity this century", is stronger outside the capital city and among men generally.[4]

Table 8.1 Selected survey results and implications for policy legitimacy and transition management

Survey result	Issues and implications
Aggregate survey results	
Strong public approval for variants of the private car.	Policy supportive of incremental innovation has relatively strong public support, potentially complicating consensus-based vision building at all levels of TM.
Most use a car frequently, but support innovations that facilitate public transport, cycling and walking.	Indicates support for policy that is inclusive rather than exclusive of options, supporting vision building and culture change at the strategic level of TM.
Electric vehicles are seen as important but do not have the same level of broad support as biofuels.	Relative to electric vehicles, State support for Finnish sourced, second generation biofuels may garner a higher level of overall public support in the short to medium term, disfavouring TM processes supportive of electric transport.
Public investment in integrated ticketing for public transport and cycling is viewed as likely to make as much difference to respondents' lives as the development of more fuel efficient conventional vehicles.	Indicates support for policy that is inclusive rather than exclusive of options, supporting vision building and culture change at strategic level of TM. Perhaps offers an entry point for reducing the population's general pro-car attitudinal disposition.
Although anthropogenic climate change is accepted by the large majority (74%), 15% think that climate change is not due to human activity; another 4% think there is no climatic change; and 7% don't know.	For a quarter of the population, anthropogenic climate change may not be a convincing policy justification, causing continuing difficulty in gaining legitimacy for problem structuring at the strategic level of TM.
Group and population differences	
Geography	
Higher car use and lower public transport use in the Oulu region; stronger environmental concern in the Helsinki region.	Geography is significant, reflected both in transport practices and environmental concerns. This is likely to affect perceptions of policy legitimacy and underlines the importance of geography in TM efforts.

continued

Table 8.1 Continued

Survey result	Issues and implications
Non-winter frequency of bicycle use is highest in Oulu, then Tampere and lowest in Helsinki. But fewer people walk daily in Oulu.	Transport practices may reflect the availability of public transport infrastructure, particularly the tram, bus and metro system available in Helsinki, but also longer commuting distances in the Helsinki region. This underlines the context-specificity of the operational level of TM.
On perception of climate change, while the median for all three regions is similar, Oulu and Tampere have similar upper quartiles of respondents who believe that "the world's climate is changing, but human activity has no effect on it during this century".	As above, possibly more so outside of Helsinki.
On the bio-economy for transport, there were significant regional differences for all but one question. Helsinki respondents are sceptical of the sustainability of current forest utilisation, while Oulu is significantly more supportive of using more of the national timber stock for biofuel.	As above, this raises issues of centre–periphery differences that merit further investigation. Also, the region surrounding Oulu is more rural than in Helsinki and Tampere, perhaps indicating a closer economic link to the use of bio-resources. Region-specific design of public engagement processes and experiments linked to TM may (but equally may not) facilitate perceived policy legitimacy.
Gender	
Significantly more men than women say that they own a car, but gender distributions of having a permanent right to use a car do not differ. However, men make disproportionate use of car travel.	May indicate differing ownership attitudes and hence support for policy affecting car use. This links to the pro-car attitudes that TM must deal with in the transition to other transport modes.
Men are more climate-sceptic, being doubtful about either the effects or actuality of anthropogenic climate change. Fewer men than women strongly agree that car use has a serious effect on climate change and fewer men strongly agree that traffic congestion in towns and cities is a very serious problem.	May affect policy legitimacy and likely to affect male response to climate messaging.

Women are significantly more likely to register "Don't know" as a response to whether the State should prioritise biofuel research above other transport technology options, and to related questions on the use of timber for biofuel production.	Policy legitimacy requires improved communication and information provision (note that this may or may not enhance public policy support).

Age

The youngest and oldest groups cycle most; the youngest group (15–24) walks the most.	Likely in part to reflect differential access to vehicles; differing practices have implications for the distribution of policy impacts and legitimacy. Yet the relevance of age and life stage is rarely discussed in TM contexts.
The youngest group (15–24) are most averse to prioritising biofuel research above other transport policy and technology options.	As above.

Income

There are significant income-based differences for "I own a car", "My family has a car" and "I have permanent right to the use of a car". In all cases the differences are particularly between the highest and the lowest income groups.	As under "Age" above. Note that the low and high income brackets are, respectively, 20–35k and 45k+ euro. In other words, the difference is barely a factor of two. Yet the relevance of income differentials is rarely discussed in TM contexts.
Median car usage increases with income, more so for non-winter use. People in lower income brackets are more frequent users of public transport.	As above.
People in the lower income brackets agree more strongly with the proposition that the current level of car use has a serious effect on climate change and on traffic congestion.	As above, and also implies the possibility of a link between transport practice and environmental attitude. Likely to also reflect age associations.

Source: Upham *et al.* 2015.

Notes

The 1,000 person sample is significantly older, contains more retirees, fewer students and is better educated than the census population ($p < 0.05$). As is common, this has been corrected with weighting factors. All results reported are statistically significant between $p < 0.001$ and $p < 0.05$. Different types of statistical test were used as appropriate. More details on the tests and significance levels can be found in Upham *et al.* 2015.

Judging by the results of the survey, in future policy Finnish State support for mobility service innovation generally is likely to be viewed as legitimate, but these policies should include 'soft' as well as 'hard' forms of mobility innovation (as referred to in Note 3, below). Figure 8.2 illustrates how respondents ranked transport innovations in terms of what would make the most positive difference to their lives. The survey found that there is majority support for relatively familiar options in two almost equal groups; one for car based solutions (fuel efficiency, smart traffic lights, hybrid vehicles), and the other for public transport (integrated public transport ticketing, light rail).

Given the heterogeneity in opinion – the existence of majority views notwithstanding – governments face a complicated task in communicating with their electorates so as to create policy legitimacy and build support. This can be seen with the notional example of biofuel-related messaging, in the sense of communicating biofuel policy. Similar issues can be expected to arise with other pro-environmental policy messaging where support is mixed among electorates, be this differentiation along the lines

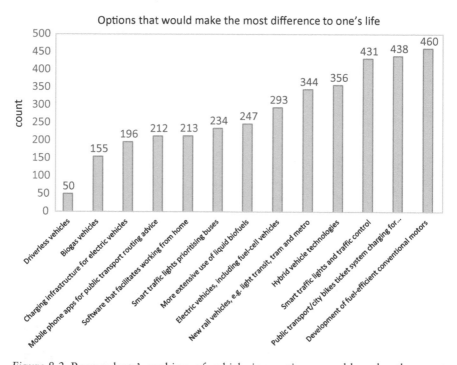

Figure 8.2 Respondents' ranking of which innovations would make the most difference to their lives.

Note
This is a response to the question: "Thinking about the options listed in Question 8, which do you think would make the most difference (positive) to your own life? Please indicate the five options that you think would make the most difference."

of age, gender, geography, social class or other dimensions. In relative terms, pro-biofuel policy related messaging can be expected to align more positively with the economic or livelihood interests of Oulu inhabitants than with those in Helsinki. General opinion towards biofuel innovation in the Helsinki region is more sceptical, particularly among the young.

As suggested above, gender opinion differences also arguably require modified modes and content of communication. Overall, women show less climate scepticism, and they perceive urban traffic congestion to be a more serious problem than men do. However, women are also more uncertain about the relative merits of different technological options for facilitating change (Upham *et al.* 2015). This implies that public information provision *may* have a greater (and perhaps positive) effect on women. While women, young, elderly and less affluent people already use low carbon transport options (i.e. public transport, cycling and walking) most frequently (all differences significant in the $p = 0.000$ to 0.003 range), among the population as a whole (as sampled), majority support is for relatively incremental innovations relating to the private car.

In the transport sector, it is implausible that the existing and emergent low carbon technological options can be deployed in time and at a sufficient scale to meet stringent decarbonisation targets (Hoen *et al.* 2009). Despite higher male support for private car innovations, there is a case for prioritising policies that strengthen the *existing* low carbon forms of transport provision. This would not exclude support for private car innovation, but it would take into account the possibilities that the demographic and geographic differences offer, and it would respond to the constrained timescale for introducing new technologies. Such an approach would also acknowledge existing path dependencies, particularly those associated with the private car (Banister *et al.* 2011; Upham *et al.* 2013). In the terminology of the MLP, low carbon technological developments associated with the private car have only recently begun to move beyond the niche level (apart from first generation biofuels). While regional variation in public opinion will complicate TM efforts at the national level, simply moving the incremental innovations beyond the niche would represent major achievements per se (Upham *et al.* 2015).

Implications: the politics of transitions management

It is clear that, compared to the generally implicit politics of innovation studies, TM has an explicit normative and political dimension, seeking as it does to support the steering of sociotechnical change in more 'sustainable' directions. While TM may have some of its roots in systems analysis, it is far from wholly analytic.

TM theory focuses on the means and modes of changing structures, cultures and practices in particular functional sectors (Frantzeskaki and de Haan 2009), transport (or mobility) being just one. Building new networks and experimenting with initiatives outside of formal institutions are commonly advocated as a means of leveraging forces operating at the niche, regime and landscape levels (Geels and Schot 2007) in mutually reinforcing directions. While until now 'publics' have not figured strongly in such analysis, the connectedness of contemporary information and communication technologies have already empowered citizens and citizen-consumers seeking change (Scammell 2000). Similarly, citizen and stakeholder consultations on new policies by public authorities, as well as topic-based polling undertaken for interest groups and others, is already common practice. What is less commonly practised, however, are more formalised ways of using consultation and poll results in policy-making and, indeed, in explicit connection with TM.

Where TM theorists have advocated learning experiments and broader network formation outside of formal institutions, they have done so primarily in recognition of the role of new networks in generating ideas, agendas and generally supporting change (Loorbach and Rotmans 2010), rather than for any normative or principled commitment to participatory democracy. In general, TM experiments seem to involve a relatively small number of selected change agents and stakeholders (e.g. Nevens *et al.* 2013; Nevens and Roorda 2014), rather than the public or publics at large. Moreover, without some deliberate policy on how the issue of democratic representativeness is to be tackled in TM, the approach is open in practice to the critique of commercial bias or capture (Hendriks 2008, 2009).

Thus, there is both a normative and a theoretical case for more explicit and systematic public engagement in TM processes, while also acknowledging that this raises difficult questions regarding how to structure engagement and consultation processes. Just as Coenen *et al.* (2012) emphasise the need to attend to geography in the context of production networks, so this is equally relevant in the context of TM and the role of public opinion in TM processes. In this regard, we would suggest that the critical and applied environmental planning literatures on infrastructure-siting controversies have much to offer by way of understanding, perspectives and practical advice. These literatures emphasise the roles of governance, institutional trust, perceived equity and place attachment (e.g. Upham and Shackley 2006; Devine-Wright 2008), all of which have strong local dimensions. Indeed, objections to technological innovations often do play out at the implementation stage, and often such opposition is local at its core.

Conclusion

TM advocates knowledge generation among actors from multiple sectors, and is intended to further sustainability norms. It aims to increase both the types and breadth of knowledge that is taken into account in policy related decision-making, and also to increase the range of actors involved

(Valkenburg and Cotella 2016). TM arenas are seen as particularly relevant and necessary in the context of problems of high uncertainty and normative ambiguity (ibid.), where existing policy design seems to fall short of effecting change at the pace required (or perceived to be required).

In this chapter, we have highlighted the merits of public input to sociotechnical innovation policy and to policies relevant for sociotechnical transitions more generally. We have observed that this is in any case implied in the socially directive nature of TM. Moreover, we have suggested that public engagement is desirable for reasons of policy legitimacy in the process of advancing sustainability transitions. Yet, the public or publics rarely figure as central actors in the sociotechnical transitions literature, and this despite their critical roles in legitimising and/or frustrating policy, supporting the NGOs that amplify and intermediate public opinion and as consumers whose billions of annual purchases make or break companies and drive economic trends.

Here, we have particularly focused on the potential policy legitimation role of publics, rather than their (our!) role as consumers, illustrating how public opinion, both in aggregate and as differentiated by demography and geography, has implications for policy legitimacy. Conceptually, we set this in the context of implications for public engagement in TM in general, drawing on the science and technology studies literature for its long-standing contribution in this regard. In so doing, we have also aimed to contribute to the transitions sub-literature that seeks to re-embed conceptualisations of transitions in their geographical contexts.

The observed heterogeneity of – and geographical variation in – public opinion, as well as the path dependencies linked to existing sociotechnical systems, pose challenges for societal transformative change. This heterogeneity among national, local and other publics also implies that in TM more attention needs to be given to geography and to specific local contexts from the outset. In this respect, surveys may be used as an indicative tool in assessing public opinion-related issues (including the likelihood of success) relating to niche experiments in particular regions. Seeking to account for public opinion as part of technology governance processes poses some fundamental (and, frankly, difficult) challenges, but public opinion matters crucially for pathways of sociotechnical change. Seeking congruence between public opinion, policy design and policy direction should help to legitimate the policy support that new technologies and social innovations need in order to compete with the current path-dependent systems.

Building meaningful public consultation into policy processes poses considerable challenges (Pidgeon *et al.* 2014), particularly given the restrictive time frames implied by stringent climate targets. Nonetheless, the results of public opinion vis-à-vis specific or more systemic sociotechnical innovations, which may include levels and types of project opposition or support, has the potential to strengthen or weaken stresses that are internal or external to the existing regime. Such public opinion many also constitute important elements of those stresses themselves. This, in turn, has implications for the transition processes of niche empowerment, reconstellation and adaptation that are required as the regime components realign to form a new regime configuration. Thus, we argue that sociotechnical transitions theorists and practitioners *need* to pay more attention to public opinion – and to the potential formative power therein – if they are to successfully catalyse new and durable policy directions for the future transitions of energy systems.

Notes

1 Source: ScienceDirect database, searched with 'transition management' in the title, abstract or keywords; English language only (date of search 17 August 2018).
2 This raises the issue of how the political and policy sciences conceive of change and the role of the public or publics therein. For an overview of the implications of five key theories of policy change for transitions processes, see Kern and Rogge (2018). It is notable, though, that none of these highlight processes of mobilisation of public opinion. The five theories are: Sabatier's advocacy coalition framework; Kingdon's multiple streams approach; Baumgartner's punctuated equilibrium theory; Hajer's discourse coalitions framework; and Pierson's policy feedback approach (ibid.).
3 Twelve different types of technology response options were offered, spanning private and public transport as well as options intended to reduce the need for transport. Broadly two thirds of people supported publicly funded research, development and deployment (RD&D) for technologies relating to new propulsion systems and fuels (fuel-efficient conventional vehicles; more extensive use of liquid biofuels; hybrid vehicle technologies; electric, biogas and fuel-cell vehicles; charging infrastructure for electric vehicles; and light rail). This level of support also extended to: integrated ticketing systems that span public transport types and bicycle carriage; telecommunications systems that facilitate home-working and public transport use; and smart traffic lights. However, only about 30 per cent of people supported public RD&D for autonomous (driverless) vehicles.
4 Nearly 74 per cent of the total sample agreed that "the world's climate is changing due to human activity this century". About 15 per cent agreed that the climate is changing but not due to human activity, some 4 per cent considered there to be no climatic change and 7 per cent did not know. Kruskal-Wallis tests with post hoc pairwise comparisons were used to compare regions and genders. Only statistically significant results are reported here.

Bibliography

Avelino, F. and Rotmans, J. 2009. "Power in transition: An interdisciplinary framework to study power in relation to structural change". *European Journal of Social Theory* 12:543–69.
Avelino, F. and Wittmayer, J. M. 2016. "Shifting power relations in sustainability transitions: A multi-actor perspective". *Journal of Environmental Policy & Planning* 18:628–49.

Banister, D. 2008. "The sustainable mobility paradigm". *Transport Policy* 15:73–80.

Banister, D., Anderton, K., Bonilla, D., Givoni M. and Schwanen T. 2011. "Transportation and the environment". *Annual Review of Environment and Resources* 36:247–70.

Barker, R. 1990. *Political Legitimacy and the State*. Oxford: Clarendon Press.

Coenen, L., Benneworth, P. and Truffer, B. 2012. "Toward a spatial perspective on sustainability transitions". *Research Policy* 41(6):968–79.

de Haan, J. and Rotmans, J. 2011. "Patterns in transitions: Understanding complex chains of change". *Technological Forecasting and Social Change* 78(1):90–102.

Delgado, A., Kjølberg, K. L. and Wickson F. 2011. "Public engagement coming of age: From theory to practice in STS encounters with nanotechnology". *Public Understanding of Science* 20(6):826–45.

Devine-Wright, P. 2008. "Reconsidering public acceptance of renewable energy technologies: A critical review". In: M. Grubb, T. Jamasb and M. G. Pollitt (eds.), *Delivering a Low Carbon Electricity System: Technologies, Economics and Policy*. Cambridge: Cambridge University Press.

EC. 2011. "Future of transport: Analytical report". Flash Eurobarometer, Directorate General for Communication, European Commission and The Gallup Organization, Brussels. Available at: http://ec.europa.eu/public_opinion/flash/fl_312_en.pdf [accessed 13 January 2015].

European Climate Foundation. 2013. "Fuelling Europe's future: How auto innovation leads to EU jobs". European Climate Foundation, Brussels. Available at: https://issuu.com/fam.robertbleeker/docs/fuelling_europe_s_future-_how_auto_ [accessed 30 November 2018].

Frantzeskaki, N. and de Haan, H. 2009. "Transitions: Two steps from theory to policy". *Futures* 41:593–606.

Frantzeskaki, N., Loorbach, D. and Meadowcroft, J. 2012. "Governing societal transitions to sustainability". *International Journal of Sustainable Development* 15(1):19–36.

Geels, F. W. 2012. "A socio-technical analysis of low-carbon transitions: Introducing the multi-level perspective into transport studies". *Journal of Transport Geography* 24:471–82. doi:10.1016/j.jtrangeo.2012.01.021.

Geels, F. W. and Schot, J. 2007. "Typology of sociotechnical transition pathways". *Research Policy* 36:399–417. doi:10.1016/j.respol.2007.01.003.

Geels, F. W. and Schot, J. 2010. "The dynamics of socio-technical transitions: A socio-technical perspective". In: J. Grin, J. Rotmans and J. Schot (eds.), *Transitions to Sustainable Development: New Directions in the Study of Long Term Transformative Change*. New York: Routledge, pp. 30–101.

Goodwin, P. and Lyons, G. 2010. "Public attitudes to transport: Interpreting the evidence". *Transportation Planning and Technology* 33(1):3–17.

Hendriks, C. M. 2008. "On inclusion and network governance: The democratic disconnect of Dutch energy transitions". *Public Administration* 86(4):1009–31.

Hendriks, C. 2009. "Policy design without democracy? Making democratic sense of transition management". *Policy Sciences* 42(4):341–68.

Hoen, A. Geurs, K., de Wilde, H., Hanschke, C. and Uyterlinde, M. 2009. "CO_2 emission reduction in transport: Confronting medium-term and long-term options for achieving climate targets in the Netherlands". Netherlands Environmental Assessment Agency and Energy Research Centre of the Netherlands,

Bilthoven and The Hague. Available at: www.rivm.nl/bibliotheek/rapporten/500076009.pdf [accessed 13 January 2015].

Kemp, R., Loorbach, D. and Rotmans, J. 2007. "Transition management as a model for managing processes of co-evolution towards sustainable development". *International Journal of Sustainable Development & World Ecology* 14(1):78–91. doi:10.1080/13504500709469709.

Kern, F. and Rogge, K. S. 2018. "Harnessing theories of the policy process for analysing the politics of sustainability transitions: A critical survey". *Environmental Innovation and Societal Transitions* 27:102–17.

Loorbach, D. 2009. "Transition management for sustainable development: A prescriptive, complexity-based governance framework". *Governance* 23:161–83. doi:10.1111/j.1468-0491.2009.01471.x.

Loorbach, D. and Rotmans, J. 2010. "The practice of transition management: Examples and lessons from four distinct cases". *Futures* 42(3):237–46.

McKinsey & Company. 2009. "Roads toward a low carbon future: Reducing CO_2 emissions from passenger vehicles in the global road transportation system". McKinsey & Company, New York. Available at: www.mckinsey.com/~/media/McKinsey/dotcom/client_service/Sustainability/PDFs/roads_toward_low_carbon_future_new.ashx [accessed 30 November 2018].

Macnaghten, P. and Chilvers, J. 2014. "The future of science governance: Publics, policies, practices". *Environment and Planning C: Politics and Space* 32(3):530–48.

Markard, J., Raven, R. and Truffer, B. 2012. "Sustainability transitions: An emerging field of research and its prospects". *Research Policy* 41:955–67.

Nahuis, R. and van Lente, H. 2008. "Where are the politics? Perspectives on democracy and technology. *Science, Technology, & Human Values* 33(5):559–81.

Nevens, F. and Roorda, C. 2014. "A climate of change: A transition approach for climate neutrality in the city of Ghent (Belgium)". *Sustainable Cities and Society* 10:112–21.

Nevens, F., Frantzeskaki, N., Gorissen, L. and Loorbach, D. 2013. "Urban transition labs: Co-creating transformative action for sustainable cities". *Journal of Cleaner Production* 50:111–22.

PE International. 2013. "Life cycle CO_2e assessment of low carbon cars 2020–2030". Low Carbon Vehicle Partnership, London. Available at: www.lowcvp.org.uk/lowcvp-viewpoint/reports.asp [accessed 30 July 2013].

Pidgeon, N., Demski, C., Butler, C., Parkhill, K. and Spence, A. 2014. "Creating a national citizen engagement process for energy policy". *Proceedings of the National Academy of Sciences* 111(Suppl 4):13606–13.

Raven, R. P. J. M., Schot, J. W. and Berkhout, F. 2012. "Space and scale in sociotechnical transitions". *Environmental Innovation and Societal Transitions* 4:63–78.

Rip, A. and Kemp, R. 1998. "Technological change". In: S. Rayner and E. Malone (eds.), *Human Choice & Climate Change, Volume 2: Resources and Technology*. Columbus, OH: Battelle Press.

Rotmans, J., Kemp, R. and Asselt, M. V. 2001. "More evolution than revolution: Transition management in public policy". *Foresight* 3(1):15–31.

Rowe, G. and Frewer, L. J. 2000. "Public participation methods: A framework for evaluation". *Science, Technology, & Human Values* 25(1):3–29.

Scammell, M. 2000. "The internet and civic engagement: The age of the citizen-consumer". *Political Communication* 17(4):351–5.

Sclove, R. E. 1995. *Democracy and Technology*. New York: The Guilford Press.

Suchman, M. C. 1995. "Managing legitimacy: Strategic and institutional approaches". *The Academy of Management Review* 20(3):571–610.

Temmes, A., Virkamäki, V., Kivimaa, P., Upham, P., Hildén, M. and Lovio, R. 2014. "Innovation policy options for sustainability transitions in Finnish transport". Tekes, Helsinki. Available at: www.researchgate.net/publication/266971688_Innovation_policy_options_for_sustainability_transitions_in_Finnish_transport.

Upham, P. and Dendler, L. 2014. "Scientists as policy actors: A study of the language of biofuel research". *Environmental Science and Policy* 47:137–47.

Upham, P. and Shackley, S. 2006. "The case of a proposed 21.5 MWe biomass gasifier in Winkleigh, Devon: Implications for governance of renewable energy planning". *Energy Policy* 34:2161–72.

Upham, P., Kivimaa, P. and Virkamäki, V. 2013. "Path dependence and expectations in transport innovation policy: The case of Finland and the UK". *Journal of Transport Geography* 32:12–22.

Upham, P., Virkamäki, V., Kivimaa, P., Hildén, M. and Wadud, Z. 2015. "Sociotechnical transitions governance and public opinion: The case of passenger transport in Finland". *Journal of Transport Geography* 46:210–19. doi:10.1016/j.jtrangeo.2015.06.024.

Valkenburg, G. and Cotella, G. 2016. "Governance of energy transitions: About inclusion and closure in complex sociotechnical problems". *Energy, Sustainability and Society* 6:20. doi:10.1186/s13705-016-0086-8.

VTPI. 2010. *Online Transport Demand Management Encyclopedia*, Victoria Transport Policy Institute, BC. Available at: www.vtpi.org/tdm/ [accessed 10 December 2018].

VTT. 2012. *Low Carbon Finland 2050* [online], VTT, Helsinki. Available at: www.vtt.fi/sites/lowcfin/en/low-carbon-finland-2050 [accessed 10 December 2018].

Wellman, B. 1978. "Public participation in transportation planning". *Traffic Quarterly* 31(4):639–56.

Wynne B. 1973. "At the limits of assessment". In: G. J. Stober and D. Schumacher (eds.), *Technology Assessment and Quality of Life*. Amsterdam: Elsevier Scientific Publishing.

9 Conclusions and research directions

Introduction

The case for bringing psychology into the study of energy transitions as addressed from sociotechnical sustainability perspectives is to some extent a case for interdisciplinarity. In the way that we have presented this in Chapter 4, it is also a case for the use of mixed methods, applied in a sequence of connected but 'bracketed' studies. The sociotechnical sustainability transitions literature already connects systems-level ideas to multiple disciplines and perspectives, particularly institutional theory (Fuenfschilling and Truffer 2014, 2016), but also governance (Bos and Brown 2012), geography (Coenen *et al.* 2012; Hansen and Coenen 2015) and others, depending on the theme or empirical topic being addressed. This variety is aided by the way in which the core concepts (sociotechnical niche, regime, landscape, trajectory, pathway) 'travel' easily across disciplinary boundaries; the way in which they can be treated as either free of, or embedded in, their epistemological and ontological roots; and the way in which their mid-level of generality helps to provide a sense of the world without over-prescribing.

Given this flexibility, we suspect that the relative neglect of psychology in the literature partly relates to: (a) the markedly differing levels of analysis in focus, with psychological processes applying mostly to the micro-level of human experience, whereas much of the sociotechnical transitions literature focuses mostly on more collective processes; and (b) the dominance of experimental methods in contemporary psychology – including in psychological approaches to sustainability studies.

Throughout this book we have sought to show that psychology is not only relevant to energy transitions as studied with sociotechnical frames, but that the differences in analytic levels of systems or structure and the individual are bridgeable. At the same time, we know that we have only scratched the surface of what is possible in these terms, having drawn primarily on data to hand, gathered over several years with colleagues also active in energy social science. There is much greater depth and coherence to be had and there is a much greater variety of psychological concepts and perspectives to be drawn upon. In this chapter, we refer to some of the options, but first we rehearse the main findings and arguments of the book.

The rationale for an integrated psychology of energy transitions

Much of this book has illustrated how, and why, psychological studies can be closely coupled with sociotechnical sustainability transitions studies, mostly in the applied domain of energy. In this section we focus primarily on summing up that rationale; in the next section we focus on further research options, particularly going forward.

In the energy-related domains discussed, major behavioural changes are often required of publics, and there is a sizeable sustainable consumption psychology literature that addresses this. To select some examples some-what arbitrarily, Abrahamse and Steg (2009) analyse how psychological and socio-demographic factors relate to energy use in households; and van den Broek *et al.* (2017) study the effects of highlighting the environmental or financial benefits, or the combined benefits of both, in environmental campaigns. Highlighting benefits that match recipients' values are found to be more persuasive than highlighting the combined environmental and financial benefits, and personal norms act as a moderator.

While such studies clearly provide valuable insights for our under-standing of human behaviour in relation to sustainable consumption (spe-cifically energy), much of this research, especially when derived from consumer psychology, tends to neglect societal influences, particularly the structural relationships between society and technology as represented in sociotechnical transition studies. Indeed, most theories in the psychology of consumption implicitly assume that a working concept of individual agency is sufficient (Jackson 2005). Hence Liu *et al.* (2017) conclude: "some [such] studies have tended to understand and predict sustainable consumption at the individual level, overlooking the social and situational factors influencing consumers' decision to act sustainably". Such studies often fail to take into account: (a) the influence of the environment on the individual (Sorrell 2015); and (b) the influence that individuals may have on that environment (co-construction).

It is our argument that the behaviour change needed for successful energy transitions will require supportive action at all levels, and this requires both truly integrated, interdisciplinary perspectives and input. The core value of bringing psychology into this mix lies particularly in support-ing an understanding of individual and related collective behaviours in energy transitions *as part of* sociotechnical systems. It is precisely this per-spective that differentiates the approach of this book from much of the more disciplinary psychological work on sustainable behaviour and sus-tainable consumption, which rather aims to understand the structure of the psychological factors involved in their own terms.

To date, while phenomena such as actor motivation and behaviour have been frequently referred to implicitly in the sociotechnical transitions literature, they are rarely theorised explicitly, using psychological concepts. Hence while sociologists of science and technology have long understood technological diffusion and adoption as processes of social embedding, in the sense of technology having and requiring supporting social institutions, the psychological processes involved have received relatively little attention in the sociotechnical transitions literature. We have suggested that this may in part reflect the ontological and methodological differences between the disciplines involved. The two models presented in Chapters 2 and 5 respectively, the energy technology acceptance framework (Upham *et al.* 2015a, see Chapter 2) and especially Stones (2005) strong structuration framework (Upham *et al.* 2018, see Chapter 5), are intended to help overcome these challenges. They do so by providing a framework for: (1) an interdisciplinary analysis of energy technology acceptance; and (2) the explicit analysis of agency–structure dynamics within sociotechnical systems. Here, the strong structuration framework (Stones 2005) is proposed with the intention of accommodating psychological perspectives.

Inspiration for the accommodation of psychological perspectives can be found in Chapters 5–8. Here, several different psychological approaches to the social embedding of new energy technology are presented, and the chapters demonstrate how these approaches could be deployed in the context of sociotechnical sustainability transitions, specifically in relation to energy.

Chapter 5 addressed the psychology of expectations, using the case of hydrogen fuel cell electric vehicles (HFCEVs) as a case study. Defining expectations as beliefs about the future, the study shows how the constructs of one of the most well-known psychological models of behaviour, the theory of planned behaviour (Ajzen and Fishbein 1980; Ajzen 2011), can help to explain why individuals act on some expectations, while other expectations remain private and unrealised. Some of the constructs and phenomena that are touched upon in this chapter include attitudes, social and personal norms, perceived behavioural control and intentions. For an account of sociotechnical change, the multi-level perspective (MLP) is used.

Chapter 6 advocates Moscovici's (2000) concept of social representations as a social psychological concept spanning psychological and sociological levels and, as such, readily lending itself to connecting psychological and sociotechnical perspectives. The chapter links social representations theory with the MLP (Geels and Schot 2007) using the case of shale gas perceptions by citizens in the UK, Germany and Poland. To this end, the chapter also illustrates the roles of psychological phenomena such as anchoring and objectification that are at the heart of the concept of social representations. It shows how these can help to explain not only the psychological aspects of some energy controversies, but also some of the social and political processes that may be involved in the embedding of different energy technologies, as well as some of the discursive tendencies and developments in social media representations of those energy technologies.

Linking to the previous chapter, Chapter 7 takes another key psychological construct – values as cognitive-emotional phenomena – and explores their relevance for exemplar grassroots innovations within the sharing economy. From a sociotechnical perspective, grassroots innovations have been viewed as operating within 'niches'. The chapter shows why values matter in this context, and it discusses the extent to which values may or may not scale up from the 'niche' to influence the 'regime'. The empirical study shows that Schwarz's psychological value scale (Schwartz 1992, 2006; Schwartz *et al.* 2012) and the sociological theory of collective enactment of values (Chen *et al.* 2013) can be combined to study the influence of values in sociotechnical transitions. In sociotechnical thinking values are regarded as slow-changing phenomena, conceptually located in the background 'landscape' of taken-for-granted assumptions. As ever, this is only of multiple possible ways to study the values.

Chapter 8 considers another domain in which psychological theories could support transition studies, namely, that of public engagement in policy co-creation; in this case, regarding low carbon transport innovation policy options. The chapter focuses mainly on the perspective of transition management (TM) (Kemp *et al.* 2007; Loorbach and Rotmans 2010) – and thus expands on the sociotechnical foundations of the book – but it also discusses how psychological surveying and public opinion elicitation techniques might be adapted to provide information about sub-national differences in public opinion for the purpose of informing policy. In addition, the chapter reveals how public perceptions are driven by sociopsychological processes, providing a basis for the application of the psychological concepts considered in Chapter 7 and the theories presented in Chapter 3, as well as those considered below.

The key purpose of this book is to make the case for the importance and relevance for energy transitions of studies of individual behaviour and behaviour change in their social and systemic context. While the insights presented in Chapters 5–8 highlight the value and beneficial use of integrated, interdisciplinary research approaches, the literature review presented in Chapter 3 shows that these are still quite uncommon and that further conceptual and empirical work connecting the micro, meso and macro levels of sociotechnical processes are needed. We are not alone in calling for this: for example, Adil and Ko (2016) refer to the need to draw on socio-psychological theory when seeking to understand the diffusion of decentralised energy systems (such as the use of consumer psychology in relation to so-called 'smart grids' operating at different scales, which often assume heightened consumer participation). Likewise, McLellan *et al.*

(2016, p. 78) highlight the importance of the influence of psychological factors on energy system transitions:

> To be able to achieve a regime which brings about a new dynamic equilibrium and stability, a niche experiment harboring (nurturing) socio-technical innovations that can lead to pathways (processes) to effective system transitions is required. However, without considering the demand side and electricity consumers' awareness, preferences and behavior, it is impossible to understand the consequences, or nature of such transition pathways.

This said, it is clearly within the areas of technology acceptance and sustainable consumption that transitions researchers mostly perceive the relevance of psychology, as also reflected in some of our own work. This focus needs to be broadened. Stones' (2005) inclusion of the psychological in its social context – applied by others to sociotechnical contexts within organisations (Greenhalgh and Stones 2010; Fjellstedt 2015) – is of broad value and is not domain-specific. Indeed, while we have barely mentioned organisational psychology, this is also of relevance to the wide array of energy-related decisions and actions that are taken (and not taken), consciously or routinely, within organisations. To some extent, this short book can only be an indicator of the possibilities for a psychology of energy transitions and of how much more there is to be discussed. If the objective of close integration between disciplines is widened and the objective of close *relevance* is adopted, then the possibilities expand enormously.

How: preconditions for integrating psychology and sociotechnical transition studies

While we argue for the value of interdisciplinary integration, a precondition for this is an in-depth understanding of the associated challenges.[1] In Chapters 2 and 3 we discussed possible reasons for the lack of integration of psychological theories in sociotechnical transitions thinking, differentiating between: (1) theoretical/ontological differences; and (2) methodological differences. The predominance of variance-based methods in contemporary social psychology studies of sustainability topics is suggested as a possible factor. Individualistic ontologies and the predominance of rational choice models in contemporary behavioural psychology are also identified as barriers to integration. To some extent, methodological bracketing side-steps these issues by allowing the juxtaposition of very different studies, but closer integration of different studies requires the use of concepts that are simultaneously social and psychological, as exemplified by our illustrative studies using social representations and also values. The social norms that we refer to in Chapter 5 would be another possible bridging concept to explore in more depth. A key research objective would then be that of tracing the interconnections between social and psychological processes, examining their connections to the structural, sociotechnical processes of energy transitions.

To this end, frameworks that help to bridge individual and socio-material structures, such as strong structuration (Stones 2005); the energy cultures framework (ECF) (Stephenson *et al.* 2010; Stephenson 2018); and the integrated, social and material framework (McMeekin and Southerton 2012) are all helpful. As Chapters 4 and 5 imply, these approaches are helpful because they fulfil particular conditions: (a) supporting disciplinary and methodological integration by (b) spanning both individual and social levels. In the following sections, we identify a number of social psychological theories and concepts that meet these interrelated criteria, thus pointing towards possibilities for further examining the role of individual-level processes in and for sociotechnical perspectives of energy system change and stasis. Again, reflecting the general tenor of the book we do not attempt to be exhaustive; the suggestions we make are to some extent also aspirational, as much pointing out what *might* be done in terms of research possibilities as bringing together what we have achieved to date.

We make these suggestions in two parts. Part 1 identifies psychological theories that have not been the focus of energy transitions research so far, but which arguably hold great potential in this regard. These theories extend beyond merely a focus on the individual; they move towards a focus on individuals (and groups) in their social surroundings. This in turn offers possibilities for connecting to the sociotechnical systems processes of energy transitions. Here we highlight in particular *social identity* perspectives and *values* research, with both providing a link between research in psychology and sociology.

Part 2 invites researchers to use the research approaches introduced in part 1 and others, and not only for the study of consumers in transitions where the main focus of research is on the individual-in-transitions as it currently lies. This section readily acknowledges and emphasises that other actors, e.g. managers, retailers or politicians, are all influenced by the socio-cognitive processes that also play a role in transitions processes, and that a research focus on the psychology as well as the sociology of such actors and processes is merited.

Part 1: broadening the range of (social) psychology theories used

The aim of social psychology is to "examine psychological processes that can be observed in all human beings and that allow for social influences on individuals" (Aronson *et al.* 2014, p. 9, our translation; see also Spreng 2017 and Allport's 1954 definition, Box 1.1). As such, social psychological research offers the potential for going beyond a focus merely on the individual. Moreover, the traditionally very broad scope of social psychological research approaches and strategies of enquiry allows for a focus on the social influences so important for much sociotechnical transitions research. This may include both the psychology of publics (groups) and consumers/ end-users (groups/individuals) (Upham *et al.* 2015b; Whitmarsh *et al.* 2015) as well as the psychology of other actors, perhaps with certain professional roles, in innovation systems (Levidow and Upham 2016, 2017). Below we suggest some options for further research.

Identity

A stream of theories that is particularly promising for further research on sociotechnical transitions focus on *identity*. George Herbert Mead (1956) was one of the first writers to point to the socially constructed nature of the self. Mead's approach to identity focuses primarily on the role of communication for the construction of the social self. He argues that "a self can arise only where there is a social process within which this self has its initiation. It [the sense of self] arises within this process" (Mead 1956, p. 42). Connected to Mead's approach is the idea that every material artefact has a symbolic meaning, derived through negotiation in social interactions. In this sense, the symbolic meaning of things is of particular relevance to consumption processes.[2] From this perspective, material goods become an important part of one's self (Jackson 2005) and this self-concept, in turn, motivates behaviour.

Box 9.1 The social symbolic self

George Herbert Mead (1934, 1956) was one of the first writers to argue for the social construction of the self. He argued that the mind and the concept of the self arise out of communication. "For Mead, the self only exists as a result of conversations of significant gestures.... the individual self can only be said to exist in relation to social conversation" (Jackson 2005, p. 71). In what has been referred to as a 'conversation of significant gestures' the individual becomes aware of participating in communication and also becomes familiar with significant symbols in written and spoken language. Social conversations thus have the function of providing mechanisms for negotiating and for internalising the values, attitudes and beliefs of a group.

Whereas Mead (1934, 1956) focuses mostly on language and communication, further research extends the notion of the 'conversation of significant gestures' to people, objects and the exchange of goods and services. Thus, material artefacts may also have social symbolic roles, changing the realm of symbols and social conversations as a whole (e.g. Zheng *et al.* 2018).

In an energy transitions context, for example, the complex relation of identity, lifestyles, symbolic meaning and artefacts/products, such as air conditioning and central heating, can be interpreted as symbols of modern life and prosperity (Nye *et al.* 2010). Likewise, cars have been shown to be status symbols in many different ways, closely connected to an individual's self-concept (Stephenson *et al.* 2010; Axsen and Kurani 2013; Gazheli *et al.* 2015). Analysing such meaning from a psychological perspective – including the symbolic meanings of more sustainable technologies (Nye *et al.* 2010) – could enlighten our understanding of the interconnections between individual and collective (social) energy transition processes, potentially building upon the large body of research from other disciplinary approaches, e.g. science and technology studies.

To analyse the influence of stakeholders' personal identity and identity construction on the individual, social and societal level, the approach of Elliott and Wattanasuwan (1998) on identity and social-symbolic consumption, for general application, would seem a promising option. Their proposed analytical framework allows for the examination of two kinds of resources (material and symbolic) and two kinds of processes (individual and social) in the construction of meaning and identity, thus effectively combining the individual (micro), social (meso) and societal (macro) levels, including structural factors. Underpinning this analysis is the self-concept of a person, here including the current self-symbolism, which may encompass relationships with material/technological artefacts and personal norms (which may be internalised and have social origins).

The meaning and the construction of meaning of these material goods is posited by Elliott and Wattanasuwan (1998) as shaped by three processes: lived experience, mediated experience and discursive elaboration. Lived experience may be direct or indirect in terms of feedback from peers – for example, in relation to cars with alternative propulsion technologies (Axsen and Kurani 2013). Mediated experience relates to the presentation of symbolic resources in multiple forms of media, e.g. different types of car. The process of discursive elaboration describes the negotiating of the symbolic meaning and the self with relevant others, e.g. friends, family and colleagues. Discursive elaboration is influenced by and influences social norms and values. Elliott and Wattanasuwan (1998) also suggest adopting the concept of social representations, a psychological approach that has already proved useful in transition studies (see Chapter 6) for the analysis of social symbolism at the group level.

In addition to studies of personal identity, including studies of the symbolic meaning of the self and social symbolism among others, the social identity perspective proposed by Schmid *et al.* (2011) may prove useful in the task of analytically linking individual or group level psychological perspectives to sociotechnical structures. A more detailed overview of this perspective is given in Box 9.2. The social identity perspective (Schmid *et al.* 2011) transcends the individual-level focus. It encompasses social identity theory (SIT) and self-categorisation theory (SCT), and rests on the assumption that people group their social world in terms of distinct social categories, e.g. by gender or profession. Such categorisations function in terms of in-groups (groups to which one belongs) and out-groups (groups to which one does not belong) and these categorisations, in turn, influence individual (and group) attitudes and behaviours. This perspective has been described as being "at the heart of social psychological theories" (Schmid *et al.* 2011, p. 211), and it may be appropriate for the study of multiple social contexts and processes involving any form of collective action (in short, any kind of intergroup relations). Such contexts may include institutions, organisations, firms, governments and consumers in aggregate or as sub-groups, all types of actors that sociotechnical sustainability transitions processes address in their theorisation.

Box 9.2 Self-categorisation and social identity: 'them versus us'

Self-categorisation theory (SCT) refers to the process of self-categorising individuals in terms of particular groups, e.g. by their profession or hobbies. While an active process of self-categorisation is posited as necessary for the group membership to become meaningful, some categories are fully self-chosen (e.g. to be fan of a football club) while others are determined more by biology (e.g. gender - although this categorisation may be seen as somewhat fluid) or by birth (e.g. nationality). In addition, the theory assumes that some form of psychological activation needs to be given for the category to be expressed in relevant attitudes or behaviours (e.g. watching one's football team play).

As defined above, self-categorisation is actually highly social, reflecting a normative social fit, defined as the perceived strengths of normative differences between in-group and out-group. According to SCT, in some social situations, given psychological activation, a process of depersonalisation may start in which group beliefs become even more important for decisions than individual beliefs, sometimes even over-riding personal identity. Thus, self-categorisation is part of: (1) social identification, although additional components are likely to be needed for the expression in attitudes and behaviours, namely: (2) evaluation (positive or negative evaluation of the group); (3) perceived importance of group affiliation (for self-identity); (4) attachment to the group and perceived interdependence between oneself and the group; (5) the degree of social embeddedness of the group in everyday life; (6) the degree of behavioural involvement in the group, defined as the extent to which behaviour functionally depends on social identity; and (7) the content and meaning of the group, including its perceived narratives (Schmid *et al.* 2011).

SCT is a sub-theory of the social identity perspective with a focus on "the social-cognitive architecture of social identity processes" (Abrams and Hogg 2010, p. 180). Social identity theory (SIT) (Tajfel and Turner 1986) deals with intergroup relations and group processes. In essence, SIT is a "theory of intergroup relations that aims to explain how individuals perceive, and act as a consequence of, their membership in social groups" (Schmid *et al.* 2011, p. 218). The theory addresses both group and intergroup processes and behaviour. Each individual may also belong to different reference groups at any one time. SIT presumes that each group member is motivated to feel good about belonging to this particular group, thus attaining or maintaining positive distinctiveness; indeed, positive social identity is a factor in keeping social groups together. This tendency can result in the often observed in-group bias, e.g. a tendency to evaluate in-group members more positively than out-group members, though it may not necessarily involve out-group derogation (Schmid *et al.* 2011).

With regard to the role of self-categorisation and social identity processes for energy transitions, the study of social movements and campaigns would seem a particularly interesting field of application. This role is addressed by Polletta and Jasper (2001) who view social identity as helping to account for the development of social movements where socio-economic class interests offer limited explanatory value – e.g. campaigns opposing nuclear power, lesbian, gay, bisexual and transgender (LGBT) rights and so on – although they caution against the over-attribution of agency to social identity in this context (ibid.). Reviewing literature addressing the question of why people join such collective movements, Polletta and Jasper (2001) identify the factors that are critical in recruiting participants as including self-interest, altruism, bonds of loyalty and solidarity, pre-existing ties and the opportunity to form new ties and activists' efforts to strategically frame identities. Pointing to the importance of frames in this regard, Polletta and Jasper (2001, p. 291) argue that

> "frames" are the interpretive packages that activists develop to mobilize potential adherents and constituents.... When successful, frames make a compelling case for the "injustice" of the condition and the likely effectiveness of collective "agency" in changing that condition. They also make clear the "identities" of the contenders, distinguishing "us" from "them" ...

In such cases, then, identity becomes both an outcome of successful strategising and a resource for it (Polletta and Jasper 2001).

In relation to low carbon mobility, an example of work deploying concepts close to those of Polletta and Jasper is Becker *et al.* (2018), who examine the way in which identity work by cycling campaigners in Berlin has improved the prospects for the institutionalisation of additional cycling protection and support measures. Other recent studies from the transition literature also support the strong influence of identity processes on the development of social movements. For example, using the case of the UK Transition Towns movement Seyfang and Haxeltine (2012) describe grass-roots movements as a means for people to express their values, their personal identity and, in part, their social identity. In further studies, social identity and self-categorisation perspectives have the potential to support the analysis of group building, seeing this as part of social movements or as a part of other groups, for example energy communities or similar – perhaps membership of free online reuse groups such as those studied in Chapter 7. Potentially, SCT could be used for the analysis of self-categorisation processes relating to such group membership, while SIT would offer an approach to the study of (especially in-group and out-group) group behaviours. Insights gained from such perspectives would, in turn, facilitate a deeper understanding of some of the relevant social psychological dynamics important for sociotechnical processes, for example social dynamics that may play a part in the strengthening of niche activities – or almost any other activity that requires cohesive social action.

Values

Value-focused approaches to topics of relevance for energy transitions, as exemplified in Chapter 7, offer another direction for further research in the psychology of energy transitions. Values are relatively long-term phenomena, and for those using the MLP framework, values tend to be conceived as located at the landscape level. It has been said that the "landscape remains something of a 'black box'" in the MLP (Whitmarsh 2012, p. 485) and perhaps this, in part, reflects the assumption of the very slow changing nature of the landscape. As already mentioned, although the role of values is acknowledged in energy transitions studies, they have been granted very little attention in this field. However, energy transitions towards more sustainable modes of energy production and consumption require changes at the landscape level and ideally in relatively short time frames. Given the potentially transformative power in and of individual and group values, this in turn lends weight to the case for studying values from a range of research perspectives.

In the social psychology literature, values are also generally regarded as slow changing.[3] Nonetheless, while the prevalent assumption is that values develop (and become fixed) during childhood and youth, recent research views values as stable but not immutable (Axsen and Kurani 2013; Welzel 2009). For example, based on qualitative work with ten households trialling a plug-in hybrid vehicle in California, Axsen and Kurani (2013) argue that *if* a household is relatively open to change in Schwartz's (1994) terms, they are also more likely to develop sustainability-oriented values. This happens: (1) if individuals' *self-concept* is open to change, even if that change is temporary; (2) if they associate sustainability with broader motivational values that are already central to their self-concept – which, in the case studied, included benevolence, universalism and self-direction in different households; and (3) if people experience positive social support for new, sustainability-oriented values within their social networks (Axsen and Kurani 2013). Similarly, Upham *et al.* (2018) discuss *change-readiness* as a key, potential mediator of willingness to accept changes to local places involving the installation of pro-environmental infrastructure (such as proximate, low carbon energy generation or storage facilities).

While research on the role of values within the sociotechnical energy and wider sustainability transitions domain has been rare (Miller *et al.* 2014), values are a core concept in much psychological work (Beckers *et al.* 2012). In general, these studies focus on the influence of individual values on behaviour. In the broad field of social psychology, the work on individual-level values and sustainable consumption, for example, continues to grow, but this work rarely contributes explicitly to transitions studies. Hence, both the theoretical and empirical implications of individual values and value change for processes such as niche development, organisational change and others remain to be theorised.

Another area of work in value studies, coming mainly from sociology and the political sciences but including psychological concepts, indicates further directions for connecting values to energy transitions studies. This concerns the study of societal values (e.g. individualism versus collectivism) and relationships to societal processes and developments, such as processes of democracy (Welzel 2009). This line of research again offers an extensive body of theory and a broad empirical database (see e.g. Welzel 2009 for an overview) on the study of value change internationally. For example, the World Value Survey has been conducted in 80 countries, often for more than 20 years.[4]

Drawing on this database, Inglehart and colleagues (e.g. Inglehart and Welzel 2007) have developed a hypothesis of the development from materialism to post-materialism. The idea reflects Maslow's (1970) hierarchy of needs and assumes that – in general – basic needs must be fulfilled (e.g. enough food, safe environment, economic security) before individuals give attention to their self-development. This is deemed to explain the rise of post-materialist values observed within many Western countries since the 1960s as economic conditions have improved. Here, post-materialistic values may manifest as, for example, an interest in self-development or in a concern for the protection of the environment.[5] Thus, Inglehart's theory of post-materialistic value change and similar studies of societal-level approaches to values are closely connected to some of the core concerns of sustainability transitions, e.g. how to change consumption patterns or even achieve a post-consumer culture, as Cohen (2013) discusses. Questions addressing the interlinkages between shared values and patterns of consumption (material and energetic) require the application of concepts that simultaneously refer to values and explain the relationship between our social (arguably sociotechnical) environment, individual values and changes in those values.

The Axsen and Kurani (2013) study referred to above is one of the few that empirically examines such a value-centred approach to energy and mobility transitions, and it does so by analysing the "potential for consumer transitions to sustainability-oriented values and behaviours" (ibid.,

p. 70). The study finds that participation in an electric car demonstration project can be a trigger for change, but that this is not necessarily sufficient; self-identity conditions are also important. The authors' small, exploratory study underlines both the challenge that values pose to sustainability transitions, the complex nature of that challenge and hence the importance of paying closer attention to key social psychological constructs within sociotechnical frames.

Part 2: rethinking the consumer focus

To date, the literature on individual agency in sociotechnical transitions has focused largely on consumers. This might be surprising, considering the fact that several key actor groups are frequently a focus of attention in sociotechnical transition frameworks. As Geels (2012, p. 417) observes: "[t]he elements in socio-technical systems are maintained, reproduced and changed by various actor groups (e.g. firms and industries, policy makers and politicians, consumers, civil society, engineers and researchers)"). Similarly, in terms of the social acceptance of new technologies, the framework presented in Chapter 2 (Upham *et al.* 2015a, building on Wüstenhagen *et al.* 2007) identifies consumers as relevant for the diffusion of new technologies or systems but, importantly, it also emphasises that acceptance of these technologies involves multiple different actors: managers, retailers, politicians to name only a few.

Other recent work also suggests that gathering insights into, for example, the psychological decision processes of actor groups beyond consumers has value for understanding transitions processes. While most studies in this field of research come from research on mobility, they do provide inspiration for further studies in the context of energy transitions. For example, with regard to cars with alternative propulsion systems, Bakker *et al.* (2014) identify barriers to the diffusion of electric cars not only on the demand side, but also on the supply side. At least until very recently, most car manufacturers in the European Union have focused on incremental improvements of the combustion engine, while cars with alternative propulsion have only been a small part of the portfolio (Geels 2012; Bakker *et al.* 2014; Gössling and Metzler 2017). Moreover, car dealers have been reluctant to support electric vehicle (EV) sales: a recent cross-country study (Zarazua de Rubens *et al.* 2018) with data from Denmark, Finland, Iceland, Norway and Spain found that car dealers misinformed consumers about EV specifications (e.g. range, charging) and that in general they were dismissive of EVs – leading to less EV sales. The study identified various reasons for this behaviour among car dealers, with lack of knowledge of EV technologies, resultant more time-consuming sales and less profitability overall being of key importance.

Other studies still imply that there might sometimes be personal, individual-level reasons for actors to work against more radical changes to elements in and of sociotechnical systems. For example, change in production systems may threaten experts' or workers' knowledge and work status (e.g. Zapata and Nieuwenhuis 2010), and thus ultimately also become a threat to the personal and social identity of those professionally involved. Some early work by Gössling (2017) on *The Psychology of the Car* highlights that, beside the factors usually mentioned to explain resistance to change (e.g. lack of governmental support and uncertainties relating to the further development of specific technologies), identity processes influence the actions of all actors in systems, thus effectively functioning as a barrier to system change. Similarly, a study on transport transitions (Gössling *et al.* 2016) found that European Union policy officers involved with transport policy were hindered not only by institutional (structural) but also individual (agency-based) barriers to change: as one policy officer suggested, "policymakers' own lifestyles may be deeply entangled in the system of aeromobility" (Gössling *et al.* 2016, p. 91). What this implies is that those whose life concept includes a dependency on air travel, or on the conventional notion of what a car is, may be psychologically compromised when faced with the prospect of significant change to those particular aspects of their lives.

Conclusions

In this short book, we have illustrated some of the ways in which insights from social psychology might be beneficial, and indeed necessary, for understanding some of the multiple concerns relating to energy transitions addressed from sociotechnical perspectives. We have not attempted to be empirically, methodologically or theoretically comprehensive, and we have been eclectic rather than mono-theoretic in our perspectives. In other words, the book sets down a marker in terms of possibilities for future research enquiry, and it invites deeper and more comprehensive treatment of the topics others (and we) touch upon. We do not argue for an individual-level (or individualistic) view of agency over and above other perspectives of agency, but rather that individual-level processes are often important in sociotechnical sustainability transitions processes. As we have said, we only scratch the surface in illustrating this and in showing how psychological and structural-level accounts can be connected. We also leave the way open for the use of different ontologies, both within and between studies.

A note of methodological caution is appropriate here: while the variety of theories referred to in this chapter can be relevant for understanding aspects of individual attitudes and behaviour, and hence contribute to a fuller account of processes of organisational decision-making and change – and thus to institutional and sociotechnical processes – the selection and application of particular theories needs to match the specific context and character of the phenomena studied. In future work we aim to follow up on some of these posited connections – conceptually and empirically – in more depth. We hope that readers concur that this is a worthwhile enterprise – alongside all the many other research directions that the expanding community of socio-technical energy and wider sustainability transitions researchers are taking – and we hope to work with some of you along the way.

Notes

1 Here we omit the socio-professional challenges of publishing interdisciplinary work, which are non-trivial given the counter-incentives for academic specialisation and its consequences in terms of disciplinary affiliations and, indeed, careers.
2 See also McCracken's (1988) meaning transfer model.
3 See, for example, Duru *et al.* (2015) on alternative agricultural production systems.
4 Dependent on the country; for more information see www.worldvaluessurvey. org.
5 For a discussion of whether such developments could be seen as a development to post-materialistic values or rather self-expression values, see e.g. Welzel 2009.

Bibliography

Abrahamse, W. and Steg, L. 2009. "How do socio-demographic and psychological factors relate to households' direct and indirect energy use and savings?". *Journal of Economic Psychology* 30:711–20.
Abrams, D. and Hogg, M. A. 2010. "Social identity, self-categorization and social influence". *European Review of Social Psychology* 1:195–228.
Adil, A. M. and Ko, Y. 2016. "Socio-technical evolution of decentralized energy systems: A critical review and implications for urban planning and policy". *Renewable and Sustainable Energy Reviews* 57:1025–37. doi:10.1016/j.rser.2015.12.079.
Ajzen, I. 2011. "The theory of planned behaviour: Reactions and reflections". *Psychology and Health* 26(9):1113–27.
Ajzen, I. and Fishbein, M. 1980. *Understanding Attitudes and Predicting Social Behavior*. Englewood Cliffs, NJ: Prentice Hall.
Aronson, E., Wilson, T. and Akert, R. 2014. *Sozialpsychologie* [Social psychology]. Hallbergmoos, Germany: Pearson Studium.
Axsen, A. and Kurani, K. S. 2013. "Developing sustainability-oriented values: Insights from households in a trial of plug-in hybrid electric vehicles". *Global Environmental Change* 23:70–80.
Bakker, S., Maat, K. and Wee, B. 2014. "Stakeholders interests, expectations, and strategies regarding the development and implementation of electric vehicles: The case of the Netherlands". *Transportation Research Part A: Policy and Practice* 66:52–64.
Batel, S., Castro, P., Devine-Wright, P. and Howarth, C. 2016. "Developing a critical agenda to understand pro-environmental actions: Contributions from social representations and social practices theories. *Wiley Interdisciplinary Reviews: Climate Change* 7(5):727–45. doi:10.1002/wcc.417.
Becker, S., Bögel, P. and Upham, P. 2018. "Transport policy change – An actor-centered analysis of Berlin's Verkehrswende project". Paper presented at the 9th International Sustainability Transitions Conference, 12–14 June 2018, Manchester.

Beckers, T., Siegers, P. and Kuntz, A. 2012. "Congruence and performance of value concepts in social research". *Survey Research Methods* 6(1):13–24.

Bhaskar, R. 2014. *The Possibility of Naturalism: A Philosophical Critique of the Contemporary Human Sciences*. Abingdon, UK: Routledge.

Bos, J. J. and Brown, R. R. 2012. "Governance experimentation and factors of success in sociotechnical transitions in the urban water sector". *Technological Forecasting and Social Change* 79:1340–53. doi:10.1016/j.techfore.2012.04.006.

Chen, K. K., Lune, H. and Queen, E. L. 2013. "How values shape and are shaped by nonprofit and voluntary organizations: The current state of the field". *Nonprofit and Voluntary Sector Quarterly* 42:856–85.

Coad, A., Daunfeldt, S-O. and Halvarsson, D. 2015. "Bursting into life: Firm growth and growth persistence by age". Ratio working paper no. 264, The Ratio Institute, Stockholm.

Coenen, L., Benneworth, P. and Truffer, B. 2012. "Toward a spatial perspective on sustainability transitions". *Research Policy* 41:968–79. doi:10.1016/j.respol.2012.02.014.

Cohen, M. J. 2013. "Collective dissonance and the transition to post-consumerism". *Futures* 52:42–51. doi:10.1016/j.futures.2013.07.001.

Devine-Wright, P. 2009. "Rethinking NIMBYism: The role of place attachment and place identity in explaining place-protective action". *Journal of Community & Applied Social Psychology* 19(6):426–41. doi:10.1002/casp. 1004.

Duru, M., Therond, O. and Fares, M. 2015. "Designing agroecological transitions; A review". *Agronomy for Sustainable Development* 35(4):1237–57. doi:10.1007/s13593-015-0318-x.

Elliott, R. and Wattanasuwan K. 1998. "Consumption and the symbolic project of the self". *European Advances in Consumer Research* 3:17–20.

Fjellstedt, L. 2015. "Examining multidimensional resistance to organizational change: A strong structuration approach". PhD dissertation, The Graduate School of Education and Human Development, The George Washington University, ProQuest LLC, Ann Arbor, MI.

Fuenfschilling, L. and Truffer, B. 2014. "The structuration of socio-technical regimes – Conceptual foundations from institutional theory". *Research Policy* 43:772–91. doi:10.1016/j.respol.2013.10.010.

Fuenfschilling, L. and Truffer, B. 2016. "The interplay of institutions, actors and technologies in socio-technical systems – An analysis of transformations in the Australian urban water sector". *Technological Forecasting and Social Change* 103:298–312. doi:10.1016/j.techfore.2015.11.023.

Gazheli, A., Antal, M. and van den Bergh, J. 2015. "The behavioral basis of policies fostering long-run transitions: Stakeholders, limited rationality and social context". *Futures* 69:14–30. doi:10.1016/j.futures.2015.03.008.

Geels, F. 2012. "A socio-technical analysis of low-carbon transitions: Introducing the multi-level perspective into transport studies". *Journal of Transport Geography* 24:471–82.

Geels, F. W. and Schot, J. 2007. "Typology of sociotechnical transition pathways". *Research Policy* 36(3):399–417.

Gössling, S. 2017. *The Psychology of the Car: Automobile Admiration, Attachment and Addiction*. Amsterdam: Elsevier.

Gössling, S. and Metzler, D. 2017. "Germany's climate policy: Facing an automobile dilemma". *Energy Policy* 105:418–28.

Gössling, S., Cohen, S. A. and Hares, A. 2016. "Inside the black box: EU policy officers' perspectives on transport and climate change mitigation". *Journal of Transport Geography* 57:83–93. doi:10.1016/j.jtrangeo.2016.10.002.

Greenhalgh, T. and Stones, R. 2010. "Theorising big IT programmes in healthcare: Strong structuration theory meets actor-network theory". *Social Science & Medicine* 70:1285–94. doi:10.1016/j.socscimed.2009.12.034.

Hansen, T. and Coenen, L. 2015. "The geography of sustainability transitions: Review, synthesis and reflections on an emergent research field". *Environmental Innovation and Societal Transitions* 17:92–109. doi:10.1016/j.eist.2014.11.001.

Hibbert, J. F., Dickinson. J. E., Gössling, S. and Curtin, S. 2013. "Identity and tourism mobility: An exploration of the attitude–behaviour gap". *Journal of Sustainable Tourism* 21(7):999–1016.

Inglehart, R. and Welzel, C. 2007. *Modernization, Cultural Change, and Democracy: The Human Development Sequence.* Zagreb: Politička kultura.

Jackson, T. 2005. "Motivating sustainable consumption: A review of evidence on consumer behaviour and behavioural change". A report to the Sustainable Development Research Network, Centre for Environmental Strategy, University of Surrey, UK.

Kemp, R., Loorbach, D. and Rotmans, J. 2007. "Transition management as a model for managing processes of co-evolution towards sustainable development". *International Journal of Sustainable Development & World Ecology* 14(1):78–91. doi:10.1080/13504500709469709.

Levidow, L. and Upham, P. 2016. "Socio-technical change linking expectations and representations: Innovating thermal treatment of municipal solid waste". *Science and Public Policy* 44(2):211–24. doi:10.1093/scipol/scw054.

Levidow, L. and Upham, P. 2017. "Linking the multi-level perspective with social representations theory: Gasifiers as a niche innovation reinforcing the energy-from-waste (EfW) regime". *Technological Forecasting and Social Change* 120:1–13. doi:10.1016/j.techfore.2017.03.028.

Liu, Y., Qu, Y., Lei, Z. and Jia, H. 2017. "Understanding the evolution of sustainable consumption research". *Sustainable Development* 25(5):414–30. doi:10.1002/sd.1671.

Loorbach, D. and Rotmans, J. 2010. "The practice of transition management: Examples and lessons from four distinct cases". *Futures* 42(3):237–46.

McCracken, G. 1988. *Culture and Consumption.* Bloomington, IN: Indiana University Press.

McLellan, B. C., Chapman, A. J. and Aoki, K. 2016. "Geography, urbanization and lock-in – Considerations for sustainable transitions to decentralized energy systems". *Journal of Cleaner Production* 128:77–96. doi:10.1016/j.jclepro.2015.12.092.

McMeekin, A. and Southerton, D. 2012. "Sustainability transitions and final consumption: Practices and socio-technical systems". *Technology Analysis and Strategic Management* 24:345–61. doi:10.1080/09537325.2012.663960.

Martínez Franco, C., Feeney, O., Quinn, M. and Hiebl, M. R. W. 2017. "Position practices of the present-day CFO: A reflection on historic roles at Guinness, 1920–1945". *Revista de Contabilidad* 20:55–62. doi:10.1016/j.rcsar.2016.04.001.

Maslow, A. H. 1970. *Motivation and Personality.* New York: Harper & Row.

Mayring, P. 2000. "Qualitative content analysis". *Forum Qualitative Sozialforschung* 1(2):20.

Mead, G. 1934. *Mind, Self and Society*. Chicago, IL: University of Chicago Press.

Mead, G. 1956. "The problem of society – How we become selves". In: G. Mead (ed.), *George Herbert Mead on Social Psychology*. Chicago, IL: University of Chicago Press, pp. 19–42.

Miller, T. R., Wiek, A., Sarewitz, D., Robinson, J., Olsson, L., Kriebel, D. and Loorbach, D. 2014. "The future of sustainability science: A solutions-oriented research agenda". *Sustainability Science* 9(2):239–46. doi:10.1007/s11625-013-0224-6.

Moscovici, S. 2000. *Social Representations*. Cambridge: Polity Press.

Nye, M., Whitmarsh, L. and Foxon, T. 2010. "Sociopsychological perspectives on the active roles of domestic actors in transition to a lower carbon electricity economy". *Environment and Planning A* 42(3):697–714. doi:10.1068/a4245.

Polletta, F and Jasper, J. M. 2001. "Collective identity and social movements". *Annual Review of Sociology* 27:283–305. doi:10.1146/annurev.soc.27.1.283.

Rogers, E. 1971/2003. *Diffusion of Innovations*. New York: Free Press.

Schmid, K., Howstone, M. and Al Ramiah, A. 2011. "Self-categorization and social identification: Making sense of us and them". In: D. Chadee (ed.), *Theories in Social Psychology*. Hopoken, NJ: Wiley-Blackwell, pp. 211–31.

Schwartz, S. H. 1992. "Universals in the content and structure of values: Theoretical advances and empirical tests in 20 countries". *Advances in Experimental Social Psychology* 25:1–65.

Schwartz, S. H. 1994. "Are there universal aspects in the structure and content of human values?". *Journal of Social Issues* 50:19–45.

Schwartz, S. H. 2006. "Les valeurs de base de la personne: Théorie, mesures et applications". *Revue Française de Sociologie* 47:929–68.

Schwartz, S. H., Cieciuch, J., Vecchione, M., Davidov, E., Fischer, R., Beierlein, C., Ramos, A., Verkasalo, M., Lönnqvist, J-E., Demirutku, K., Dirilen-Gumus, O. and Konty, M. 2012. "Refining the theory of basic individual values". *Journal of Personality and Social Psychology* 103:663–88.

Seyfang, G. and Haxeltine, A. 2012. "Growing grassroots innovations: Exploring the role of community-based initiatives in governing sustainable energy transitions". *Environment and Planning C: Government and Policy* 30(3):381–400. doi:10.1068/c10222.

Sorrell, S. 2015. "Reducing energy demand: A review of issues, challenges and approaches". *Renewable and Sustainable Energy Reviews* 47:74–82. doi:10.1016/j.rser.2015.03.002.

Sowden, L-J. and Grimmer, M. 2009. "Symbolic consumption and consumers identity: An application of social identity theory to car purchase behaviour". Paper presented at the ANZMAC conference, 30 November–2 December 2009, Melbourne.

Spreng, D. 2017. "On physics and the social in energy policy". *Energy Research & Social Science* 26:112–14. doi:10.1016/j.erss.2017.01.011.

Stephenson, J. 2018. "Sustainability cultures and energy research: An actor-centred interpretation of cultural theory". *Energy Research & Social Science* 44:242–9.

Stephenson, J., Barton, B., Carrington, G., Gnoth, D., Lawson, R. and Thorsnes, P. 2010. "Energy cultures: A framework for understanding energy behaviours". *Energy Policy* 38(10):6120–9.

Stones, R. 2005. *Structuration Theory*. New York: Palgrave Macmillan.

Tajfel, H. and Turner, J. C. 1986. "The social identity theory of inter-group behaviour". In: S. Worchel and L. W. Austin (eds.), *Psychology of Intergroup Relations*. Chicago, IL: Nelson-Hall.

Upham, P., Oltra, C. and Boso, À. 2015a. "Towards a cross-paradigmatic framework of the social acceptance of energy systems". *Energy Research & Social Science* 8:100–12. doi:10.1016/j.erss.2015.05.003.

Upham, P., Lis, A., Riesch, H. and Stankiewicz, P. 2015b. "Addressing social representations in socio-technical transitions with the case of shale gas". *Environmental Innovation and Societal Transitions* 16:120–41. doi:10.1016/j.eist. 2015.01.004.

Upham, P., Johansen, K., Bögel, P. M., Axon, S., Garard, J. and Carney, S. 2018. "Harnessing place attachment for local climate mitigation? Hypothesising connections between broadening representations of place and readiness for change". *Local Environment* 23(9):912–19. doi:10.1080/13549839.2018.1488824.

van den Broek, K., Bolderdijk, J. W. and Steg, L. 2017. "Individual differences in values determine the relative persuasiveness of biospheric, economic and combined appeals". *Journal of Environmental Psychology* 53:145–56. doi:10.1016/j. jenvp. 2017.07.009.

Welzel, C. 2009. "Werte-und Wertewandelforschung". In: V. Kaina and A. Römmele (eds.), *Politische Soziologie*. Wiesbaden: VS Verlag für Sozialwissenschaften, pp. 109–39.

Whitmarsh, L. 2012. "How useful is the multi-level perspective for transport and sustainability research?". *Journal of Transport Geography* 24:483–7. doi:10. 1016/j.jtrangeo.2012.01.022.

Whitmarsh, L., Nash, N., Upham, P., Lloyd, A., Verdon, J. P. and Kendall, J-M. 2015. "UK public perceptions of shale gas hydraulic fracturing: The role of audience, message and contextual factors on risk perceptions and policy support". *Applied Energy* 160:419–30. doi:10.1016/j.apenergy.2015.09.004.

Wüstenhagen, R., Wolsink, M. and Bürer, M. J. 2007. "Social acceptance of renewable energy innovation: An introduction to the concept". *Energy Policy* 35:2683–91. doi:10.1016/j.enpol.2006.12.001.

Zapata, C. and Nieuwenhuis, P. 2010. "Exploring innovation in the automotive industry: New technologies for cleaner cars". *Journal of Cleaner Production* 18:14–20.

Zarazua de Rubens, G., Noel, L. and Sovacool, B. K. 2018. "Dismissive and deceptive car dealerships create barriers to electric vehicle adoption at the point of sale". *Nature Energy* 3(6):501–7. doi:10.1038/s41560-018-0152-x.

Zheng, X., Baskin, E. and Peng, S. 2018. "Feeling inferior, showing off: The effect of nonmaterial social comparisons on conspicuous consumption". *Journal of Business Research* 90:196–205. doi:10.1016/j.jbusres.2018.04.041.

Appendix
List of key source papers for chapters

Where chapters draw substantially on empirics from published papers, the collection and analysis of which involved additional colleagues, these are acknowledged below.

Chapter 2

Upham, P., Oltra, C. and Boso, À. 2015. "Social acceptance of energy technologies, infrastructures and applications: Towards a general cross-paradigmatic analytical framework". *Energy Research & Social Science* 8:100–12.

Chapter 3

Bögel, P. M. and Upham, P. 2018. "Role of psychology in sociotechnical transitions studies: Review in relation to consumption and technology acceptance". *Environmental Innovation and Societal Transitions* 28:122–36. doi.10.1016/j.eist.2018.01.002.

Chapter 4

Upham, P., Dütschke, E., Schneider, U., Oltra, C., Sala, R., Lores, M., Bögel, P. and Klapper, R. 2017. "Agency and structure in a sociotechnical transition: Hydrogen fuel cells, conjunctural knowledge and structuration in Europe". *Energy Research & Social Science* 37:163–74. doi:10.1016/j.erss.2017.09.040.

Chapter 6

Upham, P., Lis, A., Riesch, H. and Stankiewicz, P. 2015. "Theorising social representations in socio-technical transitions with the case of shale gas". *Environmental Innovation and Societal Transitions* 16:120–41. doi:10.1016/j.eist.2015.01.004.

Chapter 7

Martin C. J. and Upham, P. 2016. "Grassroots social innovation and the mobilisation of values in collaborative consumption: A conceptual model". *Journal of Cleaner Production* 134:204–13. doi:10.1016/j.jclepro. 2015.04.062.

Chapter 8

Upham, P., Virkamäki, V., Kivimaa, P., Hildén, M. and Wadud, Z. 2015. "Socio-technical transitions governance and public opinion: The case of passenger transport in Finland". *Journal of Transport Geography* 46:210–19. doi:10.1016/j.jtrangeo.2015.06.024.

Index

Page numbers in **bold** denote tables, those in *italics* denote figures.

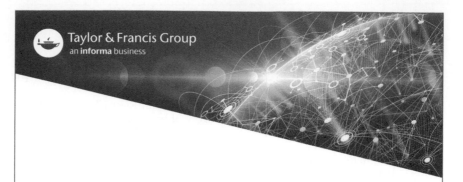

Taylor & Francis Group
an **informa** business

Taylor & Francis eBooks

www.taylorfrancis.com

A single destination for eBooks from Taylor & Francis
with increased functionality and an improved user
experience to meet the needs of our customers.

90,000+ eBooks of award-winning academic content in
Humanities, Social Science, Science, Technology, Engineering,
and Medical written by a global network of editors and authors.

TAYLOR & FRANCIS EBOOKS OFFERS:

A streamlined
experience for
our library
customers

A single point
of discovery
for all of our
eBook content

Improved
search and
discovery of
content at both
book and
chapter level

REQUEST A FREE TRIAL
support@taylorfrancis.com

Routledge
Taylor & Francis Group

CRC Press
Taylor & Francis Group